Practical Animal Handling

Pergamon Veterinary Handbook Series

Series Editor: A. T. B. Edney BA BVetMed MRCVS

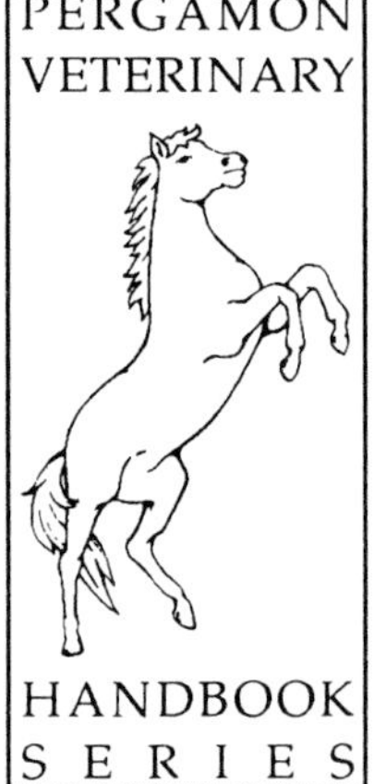

This new series of practical and authoritative handbooks covers topics of interest to the practising veterinary surgeon, to veterinary students and to veterinary nurses. The text is authoritative, yet written in a clear and accessible form, and there are numerous photographs and specially commissioned line drawings to enhance understanding. The volumes in this Series will be valuable additions to any practice bookshelf.

Other Pergamon publications of related interest

Books

EDNEY
The Waltham Book of Dog and Cat Nutrition, 2nd edition

EMILY and PENMAN
Handbook of Small Animal Dentistry

LANE
Jones's Animal Nursing, 5th edition

ROBINSON
Genetics for Cat Breeders, 2nd edition

ROBINSON
Genetics for Dog Breeders, 2nd edition

Journals

Veterinary Dermatology

The specially commissioned illustrations in this book were drawn by:
Maurizia Merati (chapters 8, 9, 10, 11, 14, 15)
Mark Iley (chapters 5, 7)
Samantha Elmhurst (chapter 12)
Jean Wheeler (chapter 6)
Ashley Waddle (chapter 16)

Practical Animal Handling

Edited by

RONALD S. ANDERSON
Department of Animal Husbandry, University of Liverpool

and

ANDREW T. B. EDNEY
Series Editor

PERGAMON PRESS
Member of Maxwell Macmillan Pergamon Publishing Corporation
OXFORD · NEW YORK · BEIJING · FRANKFURT
SÃO PAULO · SYDNEY · TOKYO · TORONTO

U.K.	Pergamon Press plc, Headington Hill Hall, Oxford OX3 0BW, England
U.S.A.	Pergamon Press, Inc., Maxwell House, Fairview Park, Elmsford, New York 10523, U.S.A.
PEOPLE'S REPUBLIC OF CHINA	Pergamon Press, Room 4037, Qianmen Hotel, Beijing, People's Republic of China
FEDERAL REPUBLIC OF GERMANY	Pergamon Press, GmbH, Hammerweg 6, D-6242 Kronberg, Federal Republic of Germany
BRAZIL	Pergamon Editora Ltda, Rua Eça de Queiros, 346, CEP 04011, Paraiso, São Paulo, Brazil
AUSTRALIA	Pergamon Press Australia Pty Ltd., P.O. Box 544, Potts Point, N.S.W. 2011, Australia
JAPAN	Pergamon Press, 5th Floor, Matsuoka Central Building, 1-7-1 Nishishinjuku, Shinjuku-ku, Tokyo 160, Japan
CANADA	Pergamon Press Canada Ltd., Suite No. 271, 253 College Street, Toronto, Ontario, Canada M5T 1R5

First edition 1991

Library of Congress Cataloging-in-Publication Data
Practical Animal Handling / edited by R. S. Anderson and A. T. B. Edney.
p. cm. — (Pergamon veterinary handbook series)
Includes index.
1. Animal handling. I. Anderson, R. S. II. Edney, A. T. B. III. Series.
SF760.A54P73 1991 636.089—dc20 90-43162

British Library Cataloguing in Publication Data
Practical Animal Handling.
1. Livestock. Husbandry
I. Anderson, R. S. (Ronald Shand) 1931– II. Edney, A. T. B.
636.089

ISBN 0-08-036151-X Hardcover
ISBN 0-08-036152-8 Flexicover

Printed in England by B.P.C.C. Wheatons Ltd, Exeter

Contents

Foreword

This book will help you to handle animals better and to advise others about how animals should be handled. Hence it is of value in three particular ways. First, the welfare of animals which are being handled will be improved. Second, as a consequence there will not be as many adverse effects on the animals; this will also benefit people, in that there will be fewer production losses or incidences of client dissatisfaction. Third, your job may become more pleasant because it will be easier to handle the animals and to carry out necessary procedures on them.

In order that animals can be successfully managed, especially when that management requires close contact, it is necessary to know about the responses of such animals to frightening or disturbing situations. The assessment of animal welfare includes the measurement of the effects on animals of feeling pain or fear, or of having any other difficulty in coping with the situation which they encounter. Animals do often have difficulties in coping with human contact and the responses to man are, to a greater or lesser extent, the responses to a dangerous predator. The biological basis of responses to potential danger is clearly explained by Alastair Lawrence in his Introduction, in which he emphasizes the extremely important role of previous experience in determining what the effect of human contact on an animal will be. If an animal is treated well and handled carefully during its development, it will be much easier to handle when it is older. This has been demonstrated by a wide range of experimental studies on laboratory and companion animals and, more recently, on farm animals.

The most valuable aspect of this book is that it contains clear descriptions of handling procedures, good drawings of these, for example, in the chapters on cattle, sheep, dogs and cats by Robert Holmes, Trevor Turner and Josephine Wills, and good photographs, for example, in the chapter on horses by Alison Schwabe.

It must be said, however, that in many cases we do not know the effects of the procedures on the animal. How many of these procedures have the required effect of holding the animal but are disturbing to it? For example, does the use of a nose loop for pigs, as described by John Walton, work because the animal is trying to minimize pain, or because it is not painful due to the action of endogenous analgesic opiates in the brain? Research should be initiated to assess the actual effects of handling on animals so that their welfare can be improved.

A further problem in animal handling is that whilst some animals have been domesticated for many thousand of years, others such as deer and mink have been kept in captivity for relatively short times. Techniques of management which are successful for species well adapted to captivity are often inappropriate for animals which retain a much higher proportion of the characteristics of their wild ancestors.

This book appears at a crossroads in time when we are moving from practical skills in handling animals, acquired and passed on as a result of many years of experience, to techniques of management based on careful scientific studies of the basic biology of the animals and of their interrelationships with man. In the future we must use this practical expertise but accelerate the rate at which we gain our understanding of the biology so that the work of those involved in animal husbandry and veterinary medicine is facilitated at the same time that the welfare of the animals is improved.

October 1990 D. M. Broom

Preface

Worldwide, a very large number of people keep a wide variety of domestic animals. Whether they do so for profit, service or pleasure, these activities carry with them a high level of responsibility for humane care.

Practically all domestic animals require some degree of handling and restraint. The scope of this interference varies from the confinement of intensive farming to the relative freedom of hill pastures or the companion animal in the home.

As a general rule, the greater the frequency of handling the less the fear or stress experienced by the animals. As a result, the risk of injury to handler and animal is also reduced. More intensive systems do not necessarily mean more handling. Automation has reduced handling so much in some cases that it may be a novel experience and thus quite stressful to many individuals. Other developments in design and husbandry, coupled with economic considerations, can have a bearing on handling techniques. Although modern farming methods may concern larger numbers of animals, there are usually fewer handlers to look after them. Changes in building design, automated feeding and cleaning, and increased physical confinement have to be made to cope with the situation. All these have important implications for animal handling and welfare. Veterinarians have a very high requirement for the handling and restraint of all domestic animals. A high percentage of these will be individuals unfamiliar to the handler.

Companion animals, including riding horses, are less influenced by economic considerations but their role has evolved a good deal in recent years from leisure interest and companionship to a wide range of work activities. These include guard, shepherd, military, tracking, detection, search and rescue, assistance and guide dogs for the deaf and blind. Other disabled people are helped by horses as well as dogs. Increased interest in saddle horses has brought with it a substantial number of new owners, many of whom have only limited experience, ability and resources for the effective management and control of their animals.

Although some degree of handling is an integral part of all animal ownership, reliable texts are not widely available. This book is aimed at all animal keepers and users, especially those who will need to gain a good deal of 'hands on' experience such as veterinary and agricultural students. The methods described are based on sound behavioural principles and the reader is particularly urged to read the opening chapter on *The Biological Basis of Handling Animals*.

No-one can be expected to become proficient in animal handling techniques simply by reading a book, any more than one can become a competent pianist or drive a car. We all need as much 'hands on' experience as we can accumulate as well. Nevertheless, the combined experience of the following chapters is very great indeed and will give readers the best possible guidance based on sound biological principles.

It is against a background of a changing environment in which animals are kept and an increasing knowledge of their behavioural responses to their environment that we have assembled the team of authors who have contributed to this book. All are experienced and knowledgeable in their field. It is an indication of their recognition of the need for a scientifically based and practically oriented book of animal handling that they were able to find the time to produce their chapters. It has been a pleasure to work with them.

We should like to pass on the thanks of our contributors to all those who have helped them with typing and proofreading of their chapters. Special thanks are due to the illustrators whose contributions have greatly enhanced the overall appearance and usefulness of this book. Finally, thank you to Mrs Marion Jowett and her colleagues at Pergamon Press for their hard work and encouragement.

October 1990

RONALD ANDERSON
ANDREW EDNEY

List of Contributors

R. S. Anderson BVMS, PhD, MRCVS: Department of Animal Husbandry, Faculty of Veterinary Science, University of Liverpool, Veterinary Field Station, Leahurst, Neston, South Wirral, Cheshire L64 7TE, United Kingdom

D. F. Broom MA, PhD, FIBiol: Professor of Animal Welfare, School of Veterinary Medicine, University of Cambridge, Madingley Road, Cambridge CB3 0ES, United Kingdom

L. A. Brown BVSc, PhD, FRCVS: Mill Leat, Gomeldon, Salisbury, Wiltshire, United Kingdom

J. E. Cooper BVSc, DTVM, MRCPath, FIBiol, FRCVS: Royal College of Surgeons of England, 35–43 Lincoln's Inn Fields, London WC2A 3PN, United Kingdom

P. E. Curtis BVSc, DipBact, MRCVS: Department of Veterinary Clinical Science, Faculty of Veterinary Science, University of Liverpool, Veterinary Field Station, Leahurst, Neston, South Wirral, Cheshire L64 7TE, United Kingdom

A. T. B. Edney BA, BVetMed, MRCVS: 22 Crocket Lane, Empingham, Rutland, LE15 8PW, United Kingdom

P. A. Flecknell MA, Vet MB, PhD, DLAS, MRCVS: Comparative Biology Centre, The Medical School, Framlington Place, Newcastle upon Tyne NE2 4HH, United Kingdom

T. J. Fletcher BVMS, PhD, MRCVS: Reediehill Deer Farm, Auchtermuchty, Cupar, Fife KY14 7HS, United Kingdom

D. M. Ford BSc, PhD: 12 Central Park, Well Head, Halifax, West Yorkshire HX1 2BT, United Kingdom

J. N. Fowler BVetMed, MRCVS: The Donkey Sanctuary, Sidmouth, Devon EX10 0NU, United Kingdom
Present address: Wiscombe Grange, South Leigh, Colyton, Devon EX13 6JS, United Kingdom

P. G. Hawkyard: Hawkyard Mink Farms Ltd, Copperas Mount, Elland, West Yorkshire, United Kingdom

R. J. Holmes BVM&S, PhD, MRCVS: Department of Veterinary Clinical Sciences, Massey University, Palmerston North 5301, New Zealand
Present address: Armadale Animal Medical Centre, 547 Dandenong Road, Armadale, Victoria 3143, Australia

A. B. Lawrence BSc, Dip Rural Sci, PhD: Animal Sciences Division, Scottish Agricultural College Edinburgh, Bush Estate, Penicuik, Midlothian EH26 0QE, United Kingdom

M. P. C. Lawton BVetMed, MRCVS: 12 Fitzilian Avenue, Harold Wood, Romford, Essex RM3 0QS, United Kingdom

A. R. Mews BVM&S, MSc, MRCVS: RSPCA, Head Office, Causeway, Horsham, West Sussex RH12 1HG, United Kingdom

A. Mowlem FIAT: The Goat Advisory Bureau, 9 Pitts Lane, Earley, Reading, Berkshire RG6 1BX, United Kingdom

A. Schwabe BSc: Department of Clinical Veterinary Medicine, Madingley Road, Cambridge CB3 0ES, United Kingdom

W. T. Turner BVetMed, MRCVS: Mandeville Veterinary Hospital, 15 Mandeville Road, Northolt, London UB5 5HD, United Kingdom

J. R. Walton BVM&S, PhD, MRCVS: Department of Veterinary Clinical Science, Faculty of Veterinary Science, University of Liverpool, Veterinary Field Station, Leahurst, Neston, South Wirral, Cheshire L64 7TE, United Kingdom

J. M. Wills BVetMed, PhD, MRCVS: Waltham Centre for Pet Nutrition, Freeby Lane, Waltham-on-the-Wolds, Leicestershire LE14 4RT, United Kingdom

1

Introduction: the biological basis of handling animals

ALISTAIR B. LAWRENCE

SINCE man first began farming animals it has been necessary for him to restrain and handle his stock. These handling events vary from procedures such as grooming and parasite control, involving a minimum of invasive techniques, to castration and the ultimate slaughter of the animal. Handling may, therefore, involve differing degrees of human presence, novelty, isolation from the social group, immobilization and pain. There is also a wide variability in animals' overt behavioural responses to handling procedures. At one extreme, unrestrained animals paradoxically show tendencies both to approach and to avoid humans (Murphey *et al.*, 1981). More acute human–animal interactions, such as the restraint by neck tether of a naive sow, provoke, in addition to vigorous avoidance, elements of aggressive behaviour (Cronin, 1985). In contrast to these 'flight and fight' responses, handling can also induce 'freezing' or immobility, the most extreme example of which is the catatonic-like state of tonic immobility found in the domestic fowl (Gallup, 1979a). Subjectively we might label these responses to restraint and handing as being indicative of 'fear' and 'stress'. From the range of animals' responses to human–animal interactions we might further intuit that, for the animal, being handled can range from a neutral, perhaps even positive, experience with a minimum of fear, to a very negative one involving much fear and stress.

It is to be expected that, within any one class of handling procedure, the behaviour of the human will play a major role in influencing fear and stress levels in the handled animal. Indirect evidence for this is shown by the variability with which pigs from a range of farms of similar size and structure are willing to approach a human observer (Hemsworth and Barnett, 1987; Fig. 1.1). If we assume

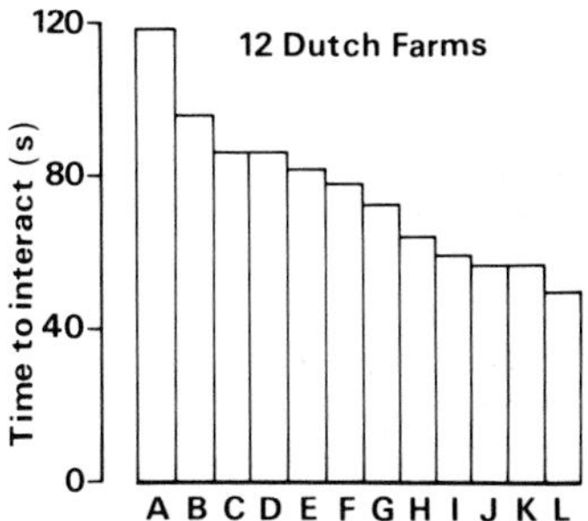

FIG. 1.1 Average time taken for pigs to interact with an experimenter in a standard test on twelve Dutch farms. (From Hemsworth and Barnett, 1987.)

that time to approach the observer represents some index of fear, these data suggest that the husbandry practised on different farms produced, in some undefined way, very different levels of fear in the animals.

In this chapter we shall examine the relationship between handling and the animals' responses. We shall consider both the external environmental factors present and those internal to the animal as influences on these responses, and consider the short- and long-term consequences of being handled on the mental and physical health of the animal.

Fear

The inference that handling procedures can induce varying degrees of fear, however intuitive it might seem, requires some justification. Fear as we know it is a negative emotional state and, being a wholly subjective experience, is not directly measurable in other members of our own species, far less in members of other species. Is there any justification for believing that animals share the experience of emotions with us?

Because of their subjective nature, emotions,

along with other unobservable mental processes, were believed for many years to be unamenable to scientific study. More recently this restricted view of animal behaviour has altered, with increasing debate on the psychobiological basis of animals' emotions (Panksepp, 1982) and the nature of animal awareness (Gallup, 1979b). This development is to some extent a result of the concern for animal welfare fuelling increased scientific debate over issues such as fear and stress in farmed livestock (Dawkins, 1980). It is more, however, a reflection of advances in animal behaviour science. The belief that the relationship between environmental stimulation and the responses of the animal has to be understood without reference to unseen mental processes has been replaced by the view that mammals and birds, like ourselves, possess cognitive abilities (Dickinson, 1980). Cognition refers to the ability of animals to acquire information about their environment by unobservable mental processes and subsequently to use this acquired information to deduce causal relationships about the world (Dunbar, 1989). From our perspective, it is important that cognitive interpretations of behaviour infer the formation of novel mental structures or images. If cognitive science tells us that animals can form mental models of the world as we do, it is not inconsistent to suppose they can also experience subjective emotions qualitively similar to our own. This view is generally upheld by the neurobehavioural evidence that emotions appear to arise from similar structures in the limbic system of the brain in humans and other animals (Panksepp, 1982).

The fear concept may be used to enable us to proceed in the analysis of animals' responses to aversive environmental stimuli without a complete understanding of the underlying physiological and neural processes. As fear may be seen as having an intermediary role in influencing the behavioural response of the animal to environmental challenge (Fig. 1.2), it is referred to as an intervening variable (Hinde, 1970). For example, the approach of humans can often induce avoidance behaviour in animals. We can say, therefore, that humans increase fear and this results in the animal fleeing. However, if the situation were as simple as this, we could exclude fear and say simply that humans increase avoidance behaviour. Where intervening variables such as fear are potentially of most use is in explaining and simplifying the complex of relationships between a number of stimuli and corresponding behavioural responses that typify the handling situation (Figs 1.2 and 1.3).

Emotions such as fear can be seen as essential in guiding the animal in its interaction with the environment. The animal is motivated to avoid stimuli that it associates with giving rise to negative emotional states, whilst it will strive to maintain contact with stimuli which it associates with positive emotional states (Weipkema, 1985; Toates, 1986).

Although we have outlined a potential role for fear in controlling behaviour, we are still unable to measure it directly. Behaviour does, however, provide a ready tool for the measurement of fear arising from handling. Rushen (1986a), for example, used the time taken to run sheep through a race as an index of their relative aversion to electro-immobilization or physical handling on the grounds that if an animal finds a handling experience unpleasant, it ought to avoid the location where it previously occurred. In pigs the length of time taken to approach an experimenter in a standard test has been used to equate with fearfulness (e.g. Hemsworth *et al.*, 1981). The use of single behavioural measures of fear is, however, compli-

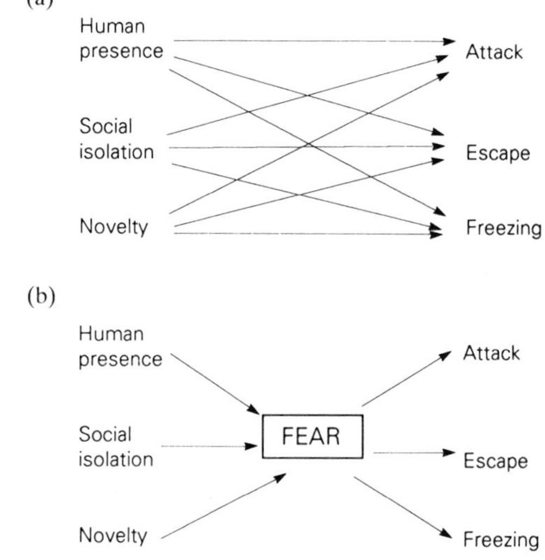

FIG. 1.2 The role of fear as an intervening variable in simplifying the relationships between stimuli and behavioural responses: (a) the three stimuli on the left can all potentially influence the three measures of fear behaviour on the right, giving nine causal relationships: (b) fear as an intervening variable reduces the number of relationships to six. (After Miller, 1959, Hinde, 1970.)

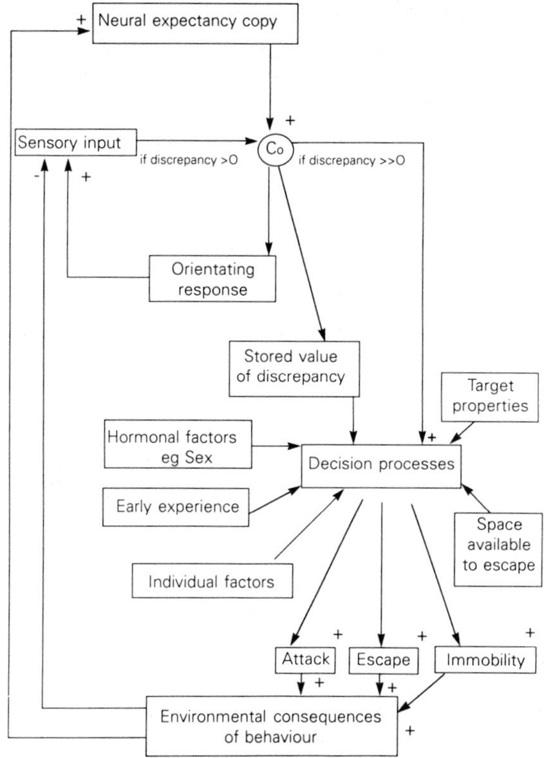

FIG. 1.3 A diagrammatic model of fear: Co refers to the neural comparator mechanism. (After Archer, 1976.)

cated by the range of behavioural responses that can be elicited by fearful stimuli (Murphy, 1978) and the absence of overt fear responses cannot always be used reliably as an indication of the absence of fear (Gray, 1987). For these reasons it is increasingly common to combine behavioural indices of fear with the physiological changes that accompany exposure to fearful stimuli.

There is a considerable overlap between the concepts of fear and stress. One distinction is that, whilst fear has previously been largely inferred from behavioural responses such as escape behaviour, stress classically refers to the responses of the autonomic and endocrine systems to a wide range of environmental challenges. Stress responses are often divided into short- and long-term reactions (see Axelrod and Reisine, 1984; Munck *et al.*, 1984 for reviews). Acute stressful stimuli or stressors activate the sympathetic–adrenal medullary system, with release of adrenaline and noradrenaline into the peripheral blood system. The ensuing changes include increases in heart rate and blood pressure, redistribution of blood from the viscera to the skeletal muscles, and mobilization of the liver's energy stores in preparation for swift action. More long-acting or chronic stressors activate the hypothalamic–pituitary–adrenal cortex axis, giving rise to raised levels of plasma glucocorticoids, with the consequent transformation of non-sugars into sugars and increased deposition of sugar in the liver. The hypothalamic–adrenocortical response thus continues the process begun in the sympathetic–adrenal medullary system by providing the animal with readily available sources of energy.

Stress, may, therefore, be regarded as an 'umbrella term' covering the relationship between environmental stimuli and the physiological changes involving sympathetic and hypothalamic–adrenocortical activation. This response to environmental challenge does not necessarily imply a physically or mentally damaging state, although it may result in somatic disorders (see p. 9). Fear,

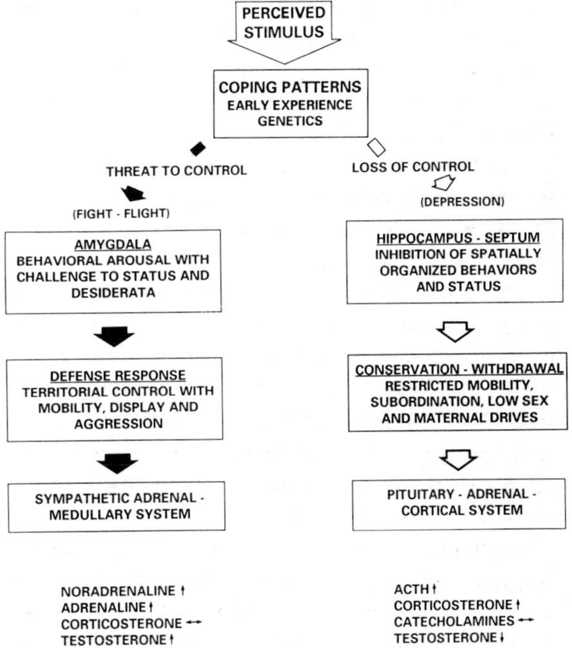

FIG. 1.4 A diagrammatic model of stress integrating fight and flight responses with immobility arising from loss of control. (From Henry, 1976, cited by Ladewig, 1987.)

the most important emotional aspect of stress, comes within the broader stress definition. Fear and stress appear most similar where the animal is exposed to a short-term challenge which elicits behavioural responses indicative of fear and also a stress activation of the sympathetic nervous system (Jones *et al.*, 1981). Fear and stress can also be closely related in situations involving long-term exposure to a challenge; for example, pigs exposed to an aversive handling treatment show increased avoidance behaviour and raised corticosteroid levels (Hemsworth *et al.*, 1986a). Fear and stress may, however, become dissociated during exposure to a long-term challenge. In tethered sows the development of stereotypic behaviour, an overt behavioural stress indicator, can in the short term be blocked by administration of anxiolytic (anxiety-reducing) drugs, but not in the long term (Odberg, 1978). Under certain chronic stress situations the animal may, therefore, experience a reduction in fear.

Behavioural Responses to Specific Aspects of Handling

It is important to consider those separate aspects of handling that may give rise to fear and stress. Some stimuli appear to be 'innately' fear-inducing and some become fear stimuli as a result of learning or conditioning. The fearful response to snakes by chimpanzees is an example of innate fear (Hebb, 1946). On the other hand, animals readily develop learned or conditioned fear responses to stimuli, harmless in themselves, which have previously been paired with fear-invoking stimuli (Hunt and Otis, 1953). Animal behaviour is, however, often the result of an amalgam between innate and learned mechanisms, as exemplified by the fear of strangers in champanzees which, although apparently developing innately, depends on prior learning in order to distinguish strange from familiar individuals (Hebb, 1946).

Special Evolutionary Dangers

Gray (1987) proposed the concept of 'special evolutionary dangers' to account for the widespread fear of predators that exists in many species, arguing that there is significant survival value in evolving fear mechanisms for the avoidance of predators. The fear responses of domestic animals for humans suggests that they perceive human beings as predators. Although domestication has brought about a dramatic reduction in the distance to which domestic animals will allow human approach before fleeing, some degree of unconditioned fear of man still clearly exists in farm animals. For example, under controlled conditions the approach of a human to a domestic hen gives rise to a graded series of responses ranging from orientation towards the distant human presence, to alarm calling and escape behaviour as the human moves closer (Jones *et al.*, 1981). The autonomic responses of the animal, as monitored by heart rate, follow a similar pattern and are consistent with the interpretation of the behavioural data that a looming human stimulus induces fear and stress in the hen. Even apparently harmless aspects of the human presence can markedly increase fear as indexed by reduced approach behaviour. The approach behaviour of pigs is inhibited when a human stands erect, wears gloves and approaches, as opposed to when he squats, shows bare hands and remains stationary (Hemsworth *et al.*, 1986b). Such aversive behavioural patterns over a period of time are associated with reduced growth rates (Gonyou *et al.*, 1986).

Another fear and stress stimulus associated with handling that has special evolutionary significance is isolation from the social group. All our domestic animals are social species and, in many respects, their social behaviour can be shown to differ little from that of their wild ancestors (e.g. Lawrence and Wood-Gush, 1988). Isolation of sheep from the social group has been found to give rise to elevated heart rate (Syme and Elphick, 1982) and plasma cortisol levels (Kilgour and DeLangen, 1970). Furthermore, in a paired choice study comparing the aversion of sheep to different handling treatments, isolation was found to be less preferred than either restraint or human presence (Rushen, 1986b). The responses of farm animals to separation from the group suggests that isolation is a significantly fearful and stressful event.

Attack as a Response to Handling

When handling animals it is obviously of interest to the handler to predict the likely response of the animal. Where the attack tendency predominates, various factors will influence whether attack proceeds and at what part of the attacker it will be aimed (Archer, 1976). The size of the attacker is one major influence on this. For example, attack behaviour in mice can be inhibited by large-sized opponents (e.g. Edwards, 1969). One way to attack a large-sized opponent such as a human is to focus on a smaller target. Not surprisingly, therefore, rats and chicks will attack the hand, whereas mice and smaller birds will attack the finger (Archer, 1976). Another factor that influences attack is movement. Generally a moving target is more effective in eliciting attack than a stationary one (Berkowitz, 1969), and in mice movement is an important aspect in determining attack behaviour (Lagerspetz, 1964).

Pain and Intense Visual, Auditory or Olfactory Stimulation

It may seem self-evident that fear can be induced by pain; for example, pigs exposed to handling treatment with a battery-operated goad showed increased avoidance behaviour and chronically elevated levels of corticosteroid (Hemsworth *et al.*, 1986a). Recent evidence, however, suggests that the central states of pain and fear are separate and that pain may not be an innate fear stimulus as was previously thought. For example, exposure of an animal to a fearful stimulus may actually reduce the pain reaction, due to fear- or stress-induced analgesia (Fanselow and Baackes, 1982). It is suggested that fear is elicited by stimuli that give anticipation of pain but not by painful stimuli themselves (Gray, 1987).

It is common knowledge that intense visual and auditory stimuli can induce fear in and affect handling of animals. The visual perception of domestic animals has been shown to influence strongly their willingness to move through races and corridors. Shadows and bright spots cause all species of livestock to baulk (Kilgour, 1976) and, therefore, the sides of races are best made solid (Grandin, 1980). Stock are, however, more likely to enter a building if it is well lit, provided that bare light bulbs emitting harsh light are avoided (Hitchcock and Hutson, 1979a). Farm animals will also show hesitation at sudden discontinuities in the floor level, and inclines generally have the effect of slowing up animal movements (Hitchcock and Hutson, 1979b). With respect to olfactory stimulation, animals can respond to odours emitted by other frightened and stressed individuals. For example, Pitman *et al.* (1988) found evidence for olfactory communication between experimental rats exposed to restraint stress and their unstressed controls. In the context of handling, the transmission of fear by odour is of obvious significance. Cattle are reported as being disturbed by the smell of blood in abattoirs or refusing to enter slaughter plants when the odours are directed from the plant towards the cattle (Grandin, 1980).

Novelty

Novelty is regarded as a potent elicitor of fear (Gray, 1987) and stress (Hennessey and Levine, 1978). In a number of the examples of fear and stress stimuli given above, novelty is an obvious confounding factor. One situation likely to induce attack or fear behaviour is the presence of an unfamiliar object in an area with which the animal is familiar (Archer, 1976). Thus, several species may direct attack or show fear toward an unfamiliar human hand (Ader, 1968; Galef, 1970). Where a chick has habituated to the human hand, attack can be reinstated by a coloured glove (Horridge, 1970, cited by Archer , 1976). Animals may also show attack or fear when placed in an unfamiliar situation (Archer, 1976). Jones and Faure (1982) showed that fear in chicks placed in novel environments increased with increases in environmental novelty and there is evidence that animals hitherto familiar with one another show an increased likelihood of attack when placed in a novel environment (Poole, 1973). Conversely, the influence of novel environments on fear behaviour can be lessened by the presence of previously familiar objects. Domestic chicks, for example, showed less fear behaviour in a novel environment if it contained something familiar to them from their rearing period (Jones, 1977).

Immobilization

Physical restraint has not been studied in the context of fear for the obvious reason that immobilization largely prevents typical fear responses and we have to assume, on circumstantial evidence, that it represents a fear stimulus. Restraint has, however, been extensively studied for its effects as a stressor (e.g. Pitman *et al.*, 1988). One much studied example of restraint is that of the tethered pregnant sow. When naive sows are first tethered, there follows a period varying from 10 to 45 minutes during which the sow throws her weight against the tether, vocalises loudly and bites the tether chain and the surrounding crate (Cronin, 1985). Over the first 2–3 weeks following tethering, sows have raised levels of plasma cortisol relative to controls (Barnett *et al.*, 1981; Becker *et al.*, 1985) and, in the long term, tethered sows show a chronic stress response relative to group-housed sows (Von Borrell and Ladewig, 1987). Whilst other factors such as feeding regimen (Appleby and Lawrence, 1987) may play a part in the development of the chronic stress response, there is little doubt that the initial period of restraint is a potent stressor and, presumably, also a fear stimulus for the sow. Restraint represents something of a special case relative to the other fear- and stress-inducing stimuli we shall consider, in that it is a clear example of an event in which the animal is largely bereft of control of its situation.

Additive Effects

The available evidence suggests that the interaction between the different types of fear and stress stimuli is additive. For example, when sheep were given a paired choice between different handling procedures, their decreasing order of preference was for human contact, physical restraint in the presence of other sheep, isolation from other sheep, capture in isolation, and inversion in isolation (Rushen, 1986b). Capture involving a combination of human presence, restraint and isolation, was preferred less than any single one of these. Furthermore, using time to travel through a race as an index of aversion, Rushen (1986a) found that restraint and electroimmobilization produced significantly greater transit times than restraint on its own. The practical implication of this is that handling procedures for domestic animals should involve as few fear- and stress-inducing stimuli as is possible.

Internal Factors Influencing the Fear State During Handling

When analysing animals' responses to handling we have to consider not only those aspects of the external environment that elicit fear and stress responses but also those internal factors that influence and control the animals' reaction to fear and stress stimuli.

Gender Effects

One of the most widely studied internal factors that influence the effects of fear stimuli is sex. The general trend from laboratory studies is for males to be the more fearful sex. For example, male chicks show more freezing and less activity, vocalization and feeding in an open field test than do females (Jones, 1977; Jones and Black, 1980). In laboratory rats, males appear more aggressive under controlled conditions (e.g. Edwards and Herndon, 1970) and also more susceptible to stress (Sawrey and Long, 1962). Where selection for fearfulness has been conducted in rats, the low fear selection line has many of the characteristics of the female sex and the high fear line those of males. For example, low fear line rats have lighter bodyweights than high fear line individuals and have more active thyroids, both female characteristics (Broadhurst, 1975; Gray, 1987). Whether these observations are applicable to behavioural responses to the handling of domestic animals has not been studied.

Sexual hormones are obvious contenders for playing a role in the development of sex differences in fear and stress responses. However, circulating levels of sex steroids can be excluded as directly responsible, since sex differences in fear behaviour persist after sex glands have been removed from both sexes (Gray, 1987). What appears to be critical is the priming of the perinatal animal at an early enough age with the appropriate hormone. Thus female rats injected with testosterone at 5 days old show marked rises in measures of fear when tested as adults (Stevens and Goldstein, 1981).

Early Experience

A major internal influence on fear and stress that has been widely examined for its effects on handling is the early or previous experience of handling itself. Regular handling of young rodents has been reported as having wide-ranging effects in adulthood such as reduced emotionality or fearfulness (Denenberg and Zarrow, 1971) and a dampened hypothalamic–adrenocortical response to novelty (Levine *et al.*, 1967). Not surprisingly, considerable interest has been shown in the effects of previous experience on the responses of domestic animals to handling. In general, regular handling of livestock has been shown to reduce fear of humans in later life. For example, regular handling of chickens from 2 days to 16 weeks reduced avoidance behaviour or flightiness (Hughes and Black, 1976). In pigs, early handling (birth to 8 weeks) and regular handling in the post-weaning period were both found to reduce fear behaviour later (Gonyou *et al.*, 1986; Hemsworth *et al.*, 1986c). In cattle it has also been shown that handled animals are less reactive in situations involving human contact (Bouisseau and Boissy, 1988). It is unclear at present to what extent handling treatments influence general fearfulness. Twice-daily handling of chicks from hatching to 6 weeks of age was found to reduce fear of humans but did not influence approaches towards an inanimate object (Jones and Faure, 1981). However, Bouisseau and Boissy (1988) found that cattle handled regularly between birth and 9 months of age also showed less apparent fear in novel situations which did not involve human contact.

It is possible that previous experience is only effective if presented at specific 'sensitive periods' in the animal's development. It is unclear at present if there is a sensitive period for previous experience of handling. Hughes and Black (1976) found that handling had no effect on avoidance behaviour in chicks if the handling regime was begun after 18 weeks, which might suggest the existence of a sensitive period prior to that time. In pigs, Hemsworth *et al.* (1986c) also found some evidence for a handling-sensitive period. However, in cattle, intensive handling limited to the first month of life was found hardly to affect calves' later behaviour (Sato *et al.*, 1984), whilst prepubertal handling extended over a long period, as mentioned earlier, did reduce fear of man (Bouissou and Boissy, 1988). This suggests that, contrary to there being a sensitive period, the most efficient strategy for improving human–animal relationships is an extended handling period stretching over much of the development of the animal.

Individual Factors

In all of the foregoing we have largely ignored the very wide intraspecific individual variability that is found in animals' responses to environmental challenges such as handling. Recent work in rodents has identified two fundamentally different ways that individual animals respond to fearful and stressful situations. On the one hand there are individuals that respond in an active manner, by fleeing or fighting, whilst on the other there are individuals that respond passively, by showing restriction of movement and immobility (Bohus *et al.*, 1987; Benus *et al.*, 1989). Similarly, in pigs it is possible to identify certain individuals that respond consistently either in an active or passive manner to a variety of non-social challenges (Lawrence *et al.*, 1989), and in goats, individuals are also found to respond to challenges in individually characteristic ways, apparently varying mainly in their degree of fearfulness (Lyons *et al.*, 1988). The importance of individual differences in determining responses to handling has not received the attention it deserves.

In the most studied example, sheep have been shown to be subdivided into individuals that are unresponsive to handling procedures and those that have been labelled the 'mindless minority baulkers' (Broad, 1977, cited by Syme and Elphick, 1982). In a study relating heart rate to behavioural responses to handling, sheep that had previously been identified as 'uncooperative' showed higher heart rates in response to handling and forced movements than those previously labelled as 'unresponsive' (Syme and Elphick, 1982). Interestingly, when sheep are driven, uncooperative individuals are also those found in the middle to back of the order, confirming the shepherds' observation that during handling the flock becomes progressively harder to control (Syme, 1981). In laying hens kept in battery cages, individuals respond to a human observer by moving to the back of the cage, by showing head-flicking behaviour, or by showing neither response (Broom *et al.*, 1986). Those individuals that move to the back of the cage or show head-flicking have relatively higher ventilation rates and plasma corticosterone responses when subsequently handled than birds that are unresponsive to being watched at close quarters. These results suggest that we can quantify individual 'temperamental' characteristics of domestic animals, such as fearfulness or stress susceptibility, which will subsequently predict the response of individuals to handling procedures. The extent to which such individual differences are genetically or environmentally determined in domestic animals is unclear. In rodents, however, selection for aggressive behaviour has also resulted in the low and high aggressive lines being, respectively, passive and active in their responses to a wide range of environmental stressors (e.g. Benus, 1988).

In summary, the internal factors that might influence responses to handling range from hormonal influences such as the sex of the individual to the effects of previous experience and individual differences. The profound effects that previous experience of humans can have on subsequent fear and stress point to the importance of this form of behavioural modification as a major means of alleviating adverse effects of handling. Further research into the factors influencing individual responses may also indicate effective methods for ameliorating fear and stress during handling.

Coping with Handling Procedures

Stimuli involved in handling situations can give rise to fear and stress by reducing the extent to which the animal can predict or control its environment (see Archer, 1976; Henry and Stephens, 1977; Wiepkema, 1985). When we consider the effects of handling procedures on the animal we are ultimately concerned to gauge the seriousness or outcome of the various fear stimuli and stressors involved. A useful guide in this is the concept of coping. Coping patterns are behavioural responses that result in the reduction of stress as measured by physiological indices (Levine *et al.*, 1978). Where orientation responses or flight and fight patterns successfully remove the effect of a fearful or stressful stimulus, these can be defined as coping patterns. This type of coping, directed at removing the stimulus responsible for the threat to control, has been called problem-focused coping (Lazarus and Folkma, 1984). Where individuals have failed to cope initially with a stressor by, for instance, escape, they may subsequently acquire coping patterns that partially or wholly reinstate physiological homeostasis even though the instigating stimulus remains. This type of coping is known as emotion-focused coping (Lazarus and Folkma, 1984). For example, the stereotyped behaviour patterns observed in tethered sows (Cronin, 1985) appear to be a form of emotion-focused coping (Dantzer and Mormede, 1983; Cronin *et al.*, 1985). Where coping breaks down, the animal is more prone to a range of somatic disorders including hypertension (Henry and Stephens, 1977), decreased immune functioning (Ballieux and Heijnen, 1987) and increased risk of tumour development (Sklar and Anisman, 1979).

There is some evidence that handling can result in physiological changes symptomatic of a breakdown in coping. For example, the fear levels of sows on different farms can be shown to be negatively correlated with reproductive performance (Hemsworth *et al.*, 1981), with herds showing a reluctance to approach a human observer also having a significantly reduced number of piglets born per sow per year. Experimentally, an unpleasant handling treatment was shown to cause increased levels of corticosterone and a reduced pregnancy rate (Hemsworth *et al.*, 1986a), suggesting that raised levels of corticosteroid are related to failure of reproductive success (e.g. Hennessy and Williamson, 1983). Unpleasant handling treatments have also been found in pigs to have a generally depressing effect on food intake (Hemsworth *et al.*, 1981; Gonyou *et al.*, 1986) although the mechanism is unclear. In chickens, unpleasant handling gives rise to decreased antibody response and resistance to *E. coli* challenge (Gross and Colmano, 1969). The welfare and the productivity of farm livestock depend on a better appreciation of how handling procedures can affect coping ability.

When an animal is first exposed to a handling treatment it will respond to this departure from the expected with species-specific behaviour to attempt to reinstate control. Under these circumstances we can expect to find, with individual exceptions (see below), physiological responses typical of flight and fight which are indicative of higher fear levels (see Fig. 1.3). However, many handling procedures involve restraint which effectively prevents the achievement of control, for example, by blocking escape behaviour. Where such blocking of fear behaviour is maintained, the animal will persist in attempting to achieve control until it perceives that it has lost control and subsequently shifts to a conservation–withdrawal state (Fig. 1.3). This model of events is upheld by a number of observations. For example, repeated immobilization is found to raise corticosterone levels in rats (Mikulaj and Mitro, 1973, cited in Henry and Stephens, 1977), and restraint of sows by neck tether also raises cortisol levels (Becker *et al.*, 1985). Sows in tethers are also less responsive to environmental stimulation, indicating that they may be in a withdrawn state (Broom, 1987). Circumstances that prevent control, either through restraint or the continued and unavoidable presence of a fearful or stressful stimulus, are clearly one way that handling can result in a loss of coping.

Loss of predictability as a result of the human–animal interaction can also cause a breakdown in coping. We have already seen that unpleasant handling can result in a reduction in growth rate. Hemsworth *et al.* (1989) have also shown that inconsistent handling can affect pig performance. Inconsistent handling, predominantly of a pleasant nature but with a small amount (20 per cent) of aversive handling, resulted in increased fear of humans, a chronic physiological stress response and a depression in growth of young pigs. Apparently trivial changes in the animals' environment may give rise to fear and ultimately loss of control. For instance, a change in the appearance of poultry stockmen (e.g. by a change from usual clothing) can result in increased fear (Jones, 1987). Therefore, handling procedures that present fearful and stressful stimuli in a variable and unpredictable way are potentially able to bring about a loss of coping.

Previous reference has been made to the variability of individual responses to handling, ranging from those which show little response to those which show an active and prolonged reaction. As yet, little is known of the respective influences of genetics and early experience on these individual behavioural styles. Regarding the constancy of these 'temperamental' characteristics, there is some initial evidence that individuals in a variety of domestic species behave in a consistent manner both across time (Kerr and Wood-Gush, 1987; Lyons *et al.*, 1988) and across different contexts (Lawrence *et al.*, 1989). Despite its importance there is little direct information available on the consequences of different individual responses to handling. However, on the basis of work in other species and in contexts other than handling, the following predictions can be made. Individuals that respond with active attack or escape behaviour appear to have a high propensity to control situations actively (e.g. Benus *et al.*, 1989). They will therefore tend to show persistent attempts to gain control and their active resistance to being handled will be correspondingly vigorous and prolonged. However, should the blocking of their attempts to control continue, they may perceive loss of control and shift to the conservation–withdrawal state. Interestingly, subjective reports from humans suggest that individuals which actively seek control view the loss of control as particularly threatening (Henry and Stephens, 1977). Subsequently, however, active responders may be more likely to develop emotion-focused coping patterns that reinstate physiological homeostasis (e.g. Dantzer *et al.*, 1988). Animals that respond passively, showing higher levels of immobility when first exposed to handling, are individuals that show a low predisposition to seek active control (e.g. Benus *et al.*, 1989). They are also likely, at least initially, to have higher levels of plasma corticosterone in response to the handling challenge (Ely, 1976, cited in Henry and Stephens, 1977) and to be unlikely in the long term to develop emotion-focused coping patterns (Dantzer *et al.*, 1988).

The potential somatic consequences of these strategies have been reviewed by Wiepkema (1985). The active response strategy will sustain an intense sympathetic activation that may lead to fatal damage of heart muscle fibres (Meerson, 1984). Stomach wall damage may also arise when the fearful and stressful stimulus (handling) ceases through rebound of the parasympathetic system after its inhibition by sympathetic activation (Glavin, 1980). For the passive strategy, a major somatic cost will be impaired immune functioning and an increased risk of disease (Ballieux and Heijnen, 1987). Both strategies may exhaust the organism if maintained for long periods (Wiepkema, 1985). It would seem that each of these alternative strategies has its own separate costs.

Alleviation of the Negative Effects of Handling

This analysis of animals' responses to handling can guide us in alleviating the negative consequences of handling. Restraint, where necessitated, should be practised for as short a period as possible, given that it so strongly blocks species-specific attempts to maintain control. Individuals without previous experience of restraint that respond actively to handling may be subject to the adverse effects of prolonged sympathetic activation. Individuals that respond without resistance are adopting an alternative strategy to attacking or escaping to gain control of the situation. This strategy may subject these individuals to deleterious physiological changes through activation of the conservation–withdrawal state and enhancement of corticosteroid output. In a situation where there is long-term blocking of problem-focused coping attempts (e.g. restraint lasting hours and days rather than minutes) passive individuals again may be more at risk through their inability to develop emotion-focused coping patterns.

Restraint is an obvious way in which man can induce a loss of coping in his livestock. Less obvious but equally important is the unpredictable presentation of fearful and stressful stimuli. The consistency with which humans handle their domestic animals may be the most critical factor in explaining the variation in fear and stress levels found on different farms (e.g. Hemsworth *et al.*, 1981). Animal handlers should therefore aim to present potentially fearful and stressful stimuli in a predictable manner that differs little from the animals' expectations of events. For example, poultry attendants often use a knock on the door as a predictor of their entry into poultry houses (B. Hughes, personal communication) and in pigs, feeding times preceded by predictable auditory cues can be shown to reduce substantially aggression and improve growth rates (Carlstead, 1986). Animals, where possible, should be handled in an area and by a human they are familiar with and, if possible, in the presence of their familiar social group.

The use of behavioural modification to decrease fear of handling through regular 'pleasant' interactions between man and his livestock should be considered and further research into the basis of individual responses might suggest alternative handling strategies for the different behavioural types. However, any 'preventive' solution to solve the problems that arise from handling which concentrates its attention on the behavioural modification of the individual, is of limited applicability to a livestock industry where some production units commonly contain thousands and tens of thousands of individuals.

Technological advances in the future may provide at least some solutions to the negative consequences of handling. Computerized 'handling' systems have the potential of allowing animals greater control over their environment. Robotic milking systems allow the animal the freedom to present itself for milking at the times it prefers and not those arbitrarily established by husbandry practice (Cross, 1983). It is expected that the future will see the development of a number of such systems that 'handle' animals semi-automatically or automatically (Wilson and Lawrence, 1987). Sow transponder feeding systems are likely to be fitted with automatic 'shedding' facilities to isolate sows for veterinary and husbandry purposes. The attraction of such systems is that, in many cases, it will be the animal, often in anticipation of some reward, that voluntarily isolates itself for veterinary treatment.

Technology can also be used to present a more predictable and less unexpected environment to the animal. The use of ear transponder tags to relay individual auditory cues allows individuals to predict when to feed at a feeding station, thereby substantially reducing aggression at the feeding site (Wierenga and Hopster, 1986). Poultry handled by an automatic broiler harvester were found to have heart rates that returned to normal more rapidly than manually handled birds and were also less fearful as measured by tonic immobility (Duncan *et al.*, 1986). In this case, automatic handling clearly presented less of a departure from the expected than the process of being caught by hand.

Summary

Handling procedures are complex, consisting of a variable number of potentially fearful and stressful stimuli. These environmental aspects of handling interact with factors internal to the animal, such as gender and previous experience, to affect its behavioural and physiological responses to being handled. Stress was taken to be an 'umbrella term' referring to the physiological activation of the animal in response to environmental challenge. Fear was regarded as the most important emotional component of stress. Both fear and stress arise as a result of the animal perceiving, through cognitive processes, that an event is in progress that differs markedly from its expectations. Perception of this discrepancy activates both the fear and stress states, resulting in fight and flight responses and raised sympathetic activity as the animal attempts to achieve control of the situation.

The negative consequences of handling can be explained by the concept of coping, which is defined as the reinstatement of physiological homeostasis following a threat to the individual's control. There is evidence to indicate that handling of farm animals can result in physiological and pathological changes indicative of a breakdown in coping. This can arise either where the animal is prevented from achieving control, for example by restraint, or where the animal is unable to predict the onset of fear and stress stimuli, such as when it is exposed to inconsistent handling.

Individual differences also affect animals' responses. Individuals that actively respond to handling by attack or escape behaviour may be at risk from the effects of the resulting prolonged sympathetic activation. Passive and cooperative individuals, however, appear to use immobility as an alternative behavioural strategy to achieve control and may be exposed to the deleterious effects of raised corticosteroid levels that result from this strategy. Restraint should be minimized and particular attention should be paid to the predictability and consistency of handling procedures. Computerized 'handling' devices may in the future alleviate some of the problems arising from handling farm animals.

References

Ader, R. (1968) Effects of early experience on emotional and physiological reactivity in the rat. *J. Comp. Physiol. Psychol.* **66,** 264–268.

Appleby, M. C. and Lawrence, A. B. (1987) Food restriction as a cause of stereotyped behaviour in tethered gilts. *Anim. Prod.* **45,** 103–110.

Archer, J. (1976) The organisation of aggression and fear in vertebrates. In: *Perspectives in Ethology, Vol. 2* (Editors Bateson, P. P. G. and Klopfer, P. H.), pp 231–298. Plenum Press, New York.

Axelrod, J. and Reisine, T. D. (1984) Stress hormones: their interaction and regulation. *Science* **224,** 452–459.

Ballieux, R. E. and Heijnen, C. J. (1987) Stress and the immune system. In: *Biology of Stress in Farm Animals: An Integrative Approach* (Editors Wiepkema, P. R. and Van Adrichem, P. W. M.), pp 29–38. Martinus Nijhoff.

Barnett, J. L., Cronin, G. M. and Winfield, C. G. (1981) The effects of individual and group penning of pigs on total and free plasma corticosteroids and the maximum corticosteroid binding capacity. *Gen. Comp. Endocrinol.* **44,** 219–225.

Becker, B. A., Ford, J. J., Christenson, R. K., Manak, R. C., Hahn, G. L. and DeShazer, J. A. (1985) Cortisol response of gilts in tether stalls. *J. Anim. Sci.* **60,** 264–270.

Benus, R. F. (1988) *Aggression and Coping.* PhD Thesis, University of Groningen, Netherlands.

Benus, R. F., Bohus, B., Koolhaas, J. M. and Van Oortmerssen, G. A. (1990) Behavioural strategies of aggressive and non-aggressive male mice in response to inescapable shock. *Behav. Processes* in press.

Berkowitz. L. (1969). The frustration-aggression hypothesis revisited. In: *Roots of Aggression* (Editor Berkowitz, L.). Atherton, New York.

Bohus, B., Benus, R. F., Fokkema, D. S., Koolhaas, J. M., Nyakas, C., Van Oortmerssen, G. A., Prins, A. J. A., de Ruiter, A. J. H., Scheurink, A. J. W. and Steffens, A. B. (1987). Neuroendocrine states and behavioural and physiological stress responses. In: *Progress in Brain Research, Vol 72, Neuropeptides and Brain Function* (Editors de Kloet, E. R., Wiegant, V. M. and de Wied, D.), pp 57–70. Elsevier, Amsterdam.

Bouisseau, M. F. and Boissy, A. (1988) Effects of early handling on heifers subsequent reactivity to humans and unfamiliar situations. In: *Proceedings of the International Congress on Applied Ethology in Farm Animals* (Editors Unshelm, J., Van Putten, G., Zeeb, K. and Ekesbo, I.), pp 21–38. KTBL, Dormstadt, Germany.

Broadhurst, P. L. (1975) The Maudsley reactive and nonreactive strains of rats: a survey. *Behav. Gen.* **5,** 299–319.

Broom, D. M. (1987) Applications of neurobiological studies to farm animal welfare. In: *Biology of Stress in Farm Animals: An Integrative Approach* (Editors Wiepkema, P. R. and Van Adrichem, P. W. M.), pp 101–111. Martinus Nijhoff.

Broom, D. M., Knight, P. G. and Stansfield, S. C. (1986) Hen behaviour and hypothalamic-pituitary-adrenal responses to handling and transport. *Appl. Anim. Behav. Sci.* **16,** 98.

Carlstead, K. (1986) Predictability of feeding: its effects on agonistic behaviour and growth in grower pigs. *Appl. Anim. Behav. Sci.* **16,** 25–38.

Cronin, G. M. (1985) *The Development and Significance of Abnormal Stereotyped Behaviour in Tethered Sows.* PhD Thesis, University of Wageningen, Netherlands.

Cronin, G. M., Wiepkema, P. R. and Van Ree, J. M. (1985) Endogenous opioids are involved in abnormal stereotyped behaviours of tethered sows. *Neuropeptides* **6,** 527–530.

Cross. M. (1983) Down on the automatic farm. *New Scientist* **108,** 56–60.

Dantzer, R. and Mormede, P. (1983) De-arousal properties of stereotyped behaviour: evidence from pituitary-adrenal correlates in pigs. *Appl. Anim. Ethol.* **10,** 233–244.

Dantzer, R., Terlouw, C., Tazi, A., Koolhaas, J. M., Bohus, B., Koob, G. F. and Le Moal, M. (1988) The propensity for schedule-induced polydipsia is related to difference in conditioned avoidance behaviour and in defense reactions in a defeat test. *Physiol. Behav.* **43,** 269–273.

Dawkins, M. S. (1980) *Animal Suffering: the Science of Animal Welfare.* Chapman and Hall, London.

Denenberg, V. H. and Zarrow, M. X. (1971) Effects of handling in infancy upon adult behaviour and adrenocortical activity: Suggestions for a neuroendocrine mechanism. In: *The Development of Self-Regulatory Mechanisms* (Editors Walcher, D. N. and Peters, D. L.), pp 39–64. Academic Press, New York.

Dickinson, A. (1980) *Contemporary Animal Learning Theory.* Cambridge University Press, Cambridge.

Dunbar, R. (1989) Common ground for thought. *New Scientist* **1646,** 48–50.

Duncan, I. J. H., Slee, G. S., Kettlewell, P., Berry, P. and Carlisle, A. J. (1986) Comparison of the stressfulness of harvesting broiler chickens by machine and by hand. *Brit. Poultry Sci.* **27,** 109–114.

Edwards, D. A. (1969) Early androgen stimulation and aggressive behaviour in male and female mice. *Physiol. Behav.* **4,** 333–338.

Edwards, D. A. and Herndon, J. (1970) Neonatal estrogen stimulation and aggressive behaviour in female mice. *Physiol. Behav.* **5,** 993–995.

Fanselow, M. S. and Baackes, M. P. (1982) Conditioned fear-induced opiate analgesia on the formalin test: evidence for two aversive motivational systems. *Learning and Motivation* **13,** 200–221.

Galef, B. G. (1970) Aggression and Timidity: Responses to novelty in feral Norway rats. *J. Comp. Physiol. Psychol.* **70,** 370–381.

Gallup, G. G. (1979a) Tonic immobility as a measure of fear in domestic fowl. *Anim. Behav.* **27,** 316–317.

Gallup, G. G. (1979b) Self awareness in primates. *Amer. Sci.* **67,** 417–421.

Glavin, G. B. (1980) Restraint ulcer: history, current research and future implications. *Brain Res. Bull.* **5,** 51–58.

Gonyou, H. W., Hemsworth, P. H. and Barnett, J. L. (1986) Effects of frequent interactions with humans on growing pigs. *Appl. Anim. Behav. Sci.* **16,** 269–278.

Grandin, T. (1980) Observations of cattle behaviour applied to the design of cattle-handling facilities. *Appl. Anim. Ethol.* **6,** 19–31.

Gray, J. A. (1987) *The Psychology of Fear and Stress.* Cambridge University Press, Cambridge.

Gross, W. B. and Colmano, G. (1969) The effect of social isolation on resistance to some infectious diseases. *Poult. Sci.* **48,** 514–520.

Hebb, D. O. (1946) On the nature of fear. *Psychological Review* **53,** 259–276.

Hemsworth, P. H. and Barnett, J. L. (1987) The human–animal relationship and its importance in pig production. *Pig News and Information* **8,** 133–136.

Hemsworth, P. H., Brand, A. and Willems, P. J. (1981) The behavioural response of sows to the presence of human beings and their productivity. *Livestock Prod. Sci.* **8,** 67–74.

Hemsworth, P. H., Barnett, J. L. and Hansen, C. (1986a) The influence of handling by humans on the behaviour, reproduction and corticosteroids of male and female pigs. *Appl. Anim. Behav. Sci.* **15,** 303–314.

Hemsworth, P. H., Gonyou, H. W. and Dziuk, P. J. (1986b) Human communication with pigs: The behavioural response of pigs to specific human signals. *Appl. Anim. Behav. Sci.* **15,** 45–54.

Hemsworth, P. H., Barnett, J. L., Hansen, C. and Gonyou, H. W. (1986c) The influence of early contact with humans on subsequent behavioural response of pigs to humans. *Appl. Anim. Behav. Sci.* **15,** 55–63.

Hemsworth, P. H., Barnett, J. L. and Hansen, C. (1990) The influence of inconsistent handling by humans on the behaviour, growth and corticosteroids of young pigs. *Appl. Anim. Behav. Sci.* in press.

Hennessey, M. B. and Levine, S. (1978) Sensitive pituitary-adrenal responsiveness to varying intensities of psychological stimulation. *Physiol. Behav.* **21,** 295–297.

Hennessy, D. P. and Williamson, P. (1983) The effects of stress and of ACTH administration in hormone profiles, oestrus and ovulation in pigs. *Theriogenology* **20,** 13–26.

Henry, J. P. and Stephens, P. M. (1977) *Stress, Health and the Social Environment. A Sociobiological Approach to Medicine.* Springer, New York.

Hinde, R. A. (1970) *Animal Behaviour.* McGraw-Hill, Tokyo.

Hitchcock, D. K. and Hutson, G. D. (1979a) Effect of variation in light intensity on sheep movement through narrow and wide races. *Aust. J. Exp. Agric. Anim. Husb.* **19,** 170–175.

Hitchcock, D. K. and Hutson, G. D. (1979b) The movement of sheep on inclines. *Aust. J. Exp. Agric. Anim. Husb.* **19,** 176–182.

Hughes, B. O. and Black, A. J. (1976) The influence of handling on egg production, egg shell quality and avoidance behaviour in hens. *Br. Poult. Sci.* **17,** 135–144.

Hunt, H. F. and Otis, L. S. (1953) Conditioned and unconditioned emotional defecation in the rat. *J. Comp. Physiol. Psychol.* **46,** 378–382.

Jones, R. B. (1977) Open-field responses of domestic chicks in the presence or absence of familiar cues. *Behav. Processes* **2,** 315–323.

Jones, R. B. (1987) The assessment of fear in the domestic fowl. In: *Cognitive Aspects of Social Behaviour in the Domestic Fowl* (Editors Zayan, R. and Duncan, I. J. H.), pp 40–81. Elsevier, Amsterdam.

Jones, R. B. and Black, A. J. (1980) Feeding behaviour of domestic chicks in a novel environment: effects of food deprivation and sex. *Behav. Processes* **5,** 173–183.

Jones, R. B. and Faure, J. M. (1981) The effects of regular handling on fear response in the domestic chick. *Behav. Processes* **6,** 135–143.

Jones, R. B. and Faure, J. M. (1982) Open-field behaviour of male and female chicks as a function of housing conditions, test situations and novelty. *Biol. Behav.* **7,** 17–25.

Jones, R. B., Duncan, I. J. H. and Hughes, B. O. (1981) The assessment of fear in domestic hens exposed to a looming human stimulus. *Behav. Processes* **6,** 121–133.

Kerr, S. G. C. and Wood-Gush, D. G. M. (1987) The development of behaviour patterns and temperament in dairy heifers. *Behav. Processes* **15,** 1–16.

Kilgour, R. (1976) *Sheep Behaviour: Its Importance in Farming Systems, Handling, Transport and Preslaughter Treatment.* West Australian Department of Agriculture, Perth, Australia.

Kilgour, R. and DeLangen, H. (1970) Stress in sheep resulting from management practices. *Proc. N. Z. Soc. Anim. Prod.* **30,** 65–76.

Ladewig, J. (1987) Endocrine aspects of stress: evaluation of stress reactions in farm animals. In: *Biology of Stress in Farm Animals: An Integrative Approach* (Editor Wiepkema, P. R. and Van Adrichem, P. W. M.), pp 13–27. Martinus Nijhoff, Dordrecht.

Lagerspetz, K. (1964) Studies on the aggressive behaviour of mice. *Ann. Acad., Sci. Fenn. Ser. B.* **131,** (3).

Lawrence, A. B. and Wood-Gush, D. G. M. (1988) Home-range behaviour and social organization of Scottish Blackface sheep. *J. Appl. Ecol.* **25,** 25–40.

Lawrence, A. B., Terlouw, E. M. C. and Illius, A. W. (1990) Analysis of individual differences in behavioural responsiveness in pigs. *Anim. Prod.* in press.

Lazarus, R. S. and Folkma, S. (1984) Coping and adaptation. In: *Handbook of Behavioural Medicine* (Editor Gentry, W. D.), pp 282–325. Guildford Press, New York.

Levine, S., Haltmeyer, G. C., Karas, G. and Denenberg, V. H. (1967) Physiological and behavioural effects on infantile stimulation. *Physiol. Behav.* **2,** 55–59.

Levine, S., Weinberg, J. and Ursin, H. (1978) Definition of the coping process and statement of the problem. In: *Psychobiology of stress: A Study of Coping Men* (Editors Ursin, H., Baade, E. and Levine, S.), pp 3–21. Academic Press, New York.

Lyons, D. M., Price, E. O. and Moberg, G. P. (1988) Individual differences in temperament of domestic dairy goats: constancy and change. *Anim. Behav.* **36,** 1323–1333.

Meerson, F. Z. (1984) *Adaptations, Stress and Prophylaxis.* Springer Verlag, Berlin.

Miller, N. E. (1959) Liberalization of basic S-R concepts: extensions to conflict behaviour, motivation and social learning. In: *Psychology: A Study of a Science, Study 1, Vol 2* (Editor Koch, S.), pp 196–299. McGraw-Hill, New York.

Munck, A., Guyre, P. M. and Holbrook, N. J. (1984) Physiological functions of glucocorticoids in stress and their relation to pharmacological actions. *End. Reviews* **5,** 25–44.

Murphey, R. M., Duarte, F. A. M. and Torres Penedo, M. C. (1981) Responses of cattle to humans in open spaces: Breed comparisons and approach-avoidance relationships. *Behav. Genet.* **11,** 37–48.

Murphy, L. B. (1978) The practical problems of recognizing and measuring fear and exploration behaviour in the domestic fowl. *Anim. Behav.* **26,** 422–431.

Odberg, F. O. (1978) Abnormal behaviours. In: *First World Congress on Ethology Applied to Zootechnics,* Madrid, pp 475–480.

Panksepp, J. (1982) Towards a general psychobiological theory of emotions. *Behav. Brain Sciences,* **5,** 406–467.

Pitman, D., Ottenweller, J. E. and Natelson, B. H. (1988) Plasma corticosterone levels during repeated presentation of two intensities of restraint stress: chronic stress and habituation. *Physiol. Behav.* **43,** 47–55.

Poole, T. (1973) The aggressive behaviour of individual polecats *Mustela putorius, M. furo* and hybrids towards familiar and unfamiliar opponents. *J. Zool.* **170,** 395–414.

Rushen, J. (1986a) Aversion of sheep to electro-immobilization and physical restraint. *Appl. Anim. Behav. Sci.* **15,** 315–324.

Rushen, J. (1986b) Aversion of sheep for handling treatments: paired-choice studies. *Appl. Anim. Behav. Sci.* **16,** 363–370.

Sato, S., Shiki, H. and Yamasaki, F. (1984) The effects of early caressing on later tractability of calves. *Jpn. J. Zootech, Sci.* **55,** 332–338.

Sawrey, W. L. and Long, D. H. (1962) Strain and sex differences in ulceration in the rat. *J. Comp. Physiol. Psychol.* **55,** 603–605.

Sklar, L. S. and Anisman, H. (1979) Stress and coping factors influence tumour growth. *Science,* **205,** 513–515.

Stevens, R. and Goldstein, R. (1981) Effects of neonatal testosterone and oestrogen on open-field behaviour in rats. *Physiol. Behav.* **26,** 551–553.

Syme, L. A. (1981) Social disruption and forced movement orders in sheep. *Anim. Behav.* **29,** 283–288.

Syme, L. A. and Elphick, G. R. (1982) Heart-rate and the behaviour of sheep in yards. *Appl. Anim. Ethol.* **9,** 31–35.

Toates, F. M. (1986) *Motivational Systems.* Cambridge University Press, Cambridge.

Von Borrell, E. and Ladewig, J. (1987) The adrenal response to chronic stress is modified by individual differences in adrenal function of pigs. *Appl. Anim. Behav. Sci.* **17,** 378.

Wiepkema, P. R. (1985) Biology of fear. In: *Second European Symposium on Poultry Welfare,* pp 84–92.

Wierenga, H. K. and Hopster, H. (1986) Behavioural research to improve systems for automatic concentrate feeding. In: *Proceedings of the International Symposium on Applied Ethology in Farm Animals,* p 18. Balantonfured, Hungary.

Wilson, P. N. and Lawrence, A. B. (1987) Changes in livestock nutrition and husbandry. In: *Farming into the Twenty-First Century* (Editor Gasser, J.), pp 119–145. Norsk Hydro, London.

2

Cattle

ROBERT J. HOLMES

Introduction

Skill and the Risk of Injury

Good cattle handling is a skill which gets a job done safely, humanely and quickly. Handling skills are particularly important to veterinarians because clients judge them on these skills. Clients usually dislike rough treatment or severe correction of their animals.

Most cattle handling can be done without physical exertion. It is best done with human ingenuity and patience — not by force. Force is met with equal or greater force. By understanding their behaviour, the actions can be predicted and thus directed to give the required response, with least likelihood of injury to animal and handler.

There is a variety of actions by cattle which can cause injury. At up to 800 kg as adults, their weight is sufficient to knock down, trample or crush a person against a wall or fence, either by intention or accidentally. Bruised or broken toes will result from trampling by cattle unless protective footwear is worn.

Cattle use their heads to attack and defend themselves. If horned, they can inflict severe damage when they attack with the head. A quick thrust sideways or forwards can be fatal. They can butt with great force and inflict injury even when polled or dehorned. Dairy bulls are particularly dangerous.

Cattle usually kick with one hind leg at a time in a forward, outwards and backwards arc (the 'cow kick') and so can strike someone standing at their shoulder. Cows are particularly protective in the first day or two after calving. They may attack anyone approaching their calves.

Although cattle do not bite offensively or defensively, and there are no incisors in the upper jaw, the molars are very effective at crushing and grinding. Serious finger and hand injuries may be inflicted during drenching of cattle.

Tail switching due to high arousal or fly attack can cause injuries to unprotected eyes when struck by the long hairs at the end of the tail.

The countermeasures that can be taken to prevent injury to handlers by cattle tend to be active rather than passive. Active countermeasures are those which require a conscious action by the person at risk, for example correct positioning to avoid getting kicked. Passive measures such as the culling and selection against aggression in beef cattle require no conscious action once they have been built into the system. Although such passive measures may reduce the risk substantially and are preferable, many risk reduction measures are combinations of active and passive. For example, an active passive measure would be putting on boots with protective toecaps. That requires an action to use a passive measure. The maintenance of raised walkways ('catwalks') alongside races, as shown in Fig. 2.1, is a passive active measure as it is a passive intervention requiring action to maintain it. At present, safety in cattle handling is almost completely dependent on active measures, with some active passive, a few passive active and hardly any passive measures. Clearly there is great scope and a need for ingenuity in designing passive countermeasures for cattle handling.

Welfare

The welfare of handled animals is important for two reasons. The first is that it is our obligation to maintain the well-being of animals in our care. The second is that the whole handling procedure is easier when the animal is relaxed.

People with dairy experience generally consider that dairy cows are easier to handle when they are

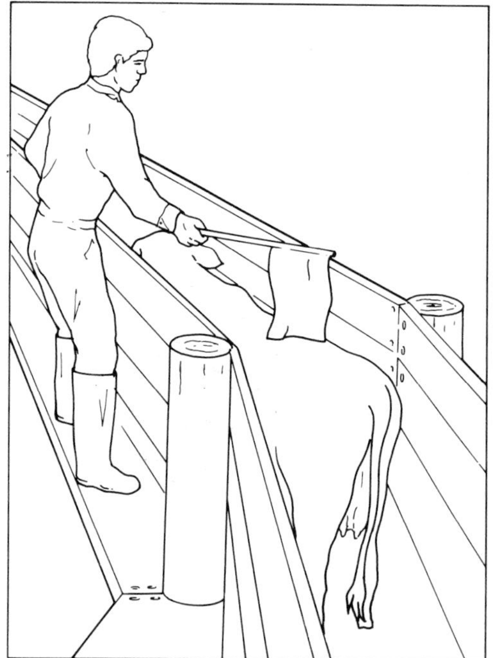

FIG. 2.1 Using a raised walkway to handle cattle in a single-file race with a flag. Note the solid sides of the race.

comfortable, i.e. when they are dry, free of draughts and clean.

Arousal

Arousal is a key concept in animal handling and is particularly relevant to the handling of large, excitable beasts like cattle. It is the state of activation of an animal and ranges from deep sleep, to fight or flight, with several states in between. The higher the arousal the more an animal reacts to stimulation. An adult dairy bull which is aroused needs very little provocation to attack. Highly aroused cattle are likely to show fight or flight and to make sudden, violent movements. When mustering in open country, it may be useful to have cattle running away, but this is undesirable in small spaces, where they may run into people or structures, jam in gaps, bruise themselves or fall over. It is important to give the minimal amount of stimulation necessary to get a response, otherwise there is a risk of getting undesirable or inappropriate behaviour such as flight.

When it cannot escape, an aroused animal which is further stimulated by handling may 'freeze'. For instance, a cow may go down in a confining pen and thus stop the flow of pregnancy testing by rectal examination. Despite stimulation she may refuse to stand for several minutes. When she does stand, it will often be without warning, which is dangerous unless anticipated.

Generally it is desirable to keep animals calm so that they move or stand quietly. Cattle or sheep should be allowed to settle down for 20 minutes or so after yarding.

Control of arousal, therefore, is fundamental to controlling the movement of animals. Handling, by its nature, stimulates animals and raises their arousal. An increase is usually undesirable and what is needed is a reduction. Some of the factors which affect arousal of cattle are shown in Fig. 2.2.

As a species which is generally more active in daylight (diurnal), cattle are naturally less active

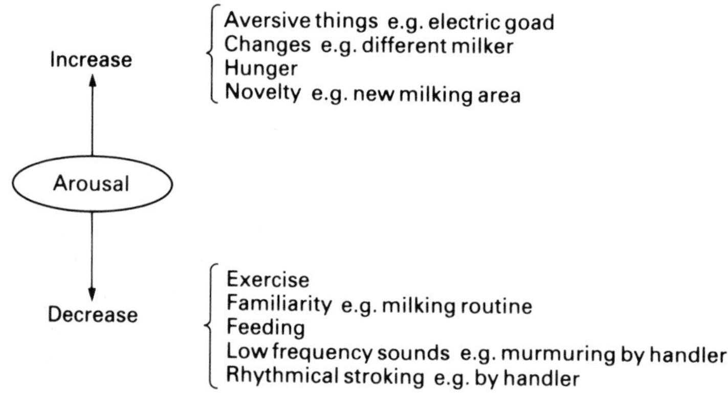

FIG. 2.2 Some factors affecting arousal of cattle. Note that the order in which factors are listed does not imply that one factor is more effective than another.

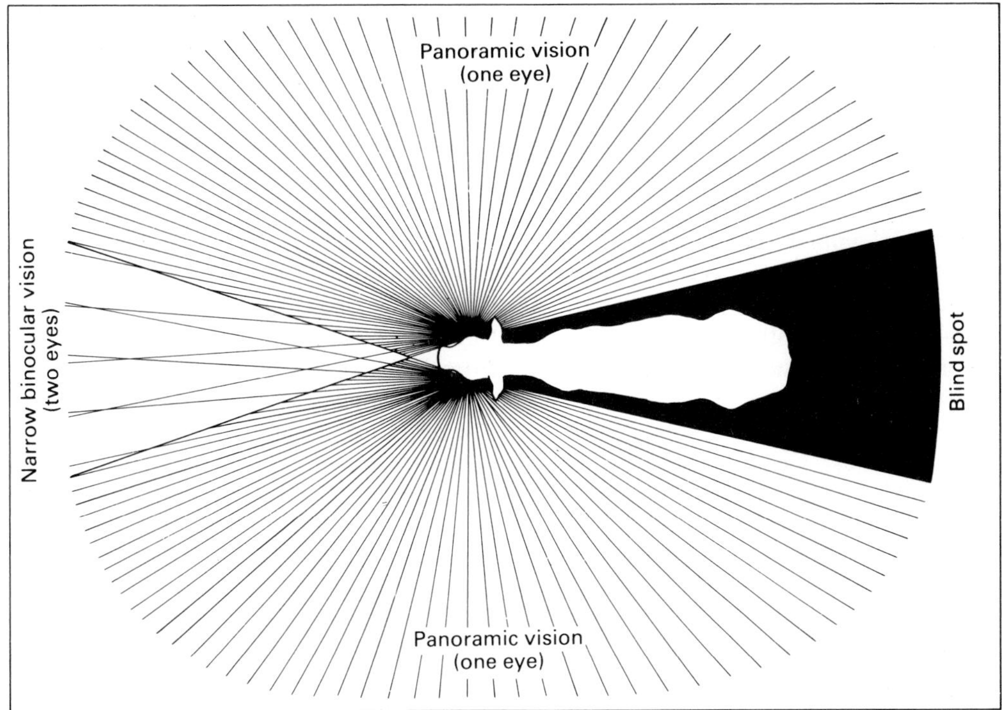

FIG. 2.3 The field of vision of a cow.

and quieter in the dark. Night time may well be a better time for handling excitable animals. Cattle can, however, be active at night, as shown by their grazing in the dark when the days are too hot.

The responsiveness of cattle varies with numerous factors. A beef cow within a day of giving birth may attack anyone approaching her calf. Newly-weaned beef cows may force their way through fences in trying to rejoin their calves. The humped cattle have a reputation for being more difficult to handle in yards. Responsiveness to stimuli is affected by an animal's temperament as well as the type and intensity of stimuli. Temperament is affected by:

—type of cattle (humped are more responsive than humpless)
—breed (Angus are more flighty than Herefords)
—strain (variation exists within breeds)
—reproductive status (newly-calved beef cows may attack people in proximity to their day-old calves)
—experience (animals habituate to things that are of no consequence, and gentle handling makes cattle quieter).

Perception

Cattle have a far more acute sense of smell than humans. Cows can smell their newly weaned calves from considerable distances, especially downwind, and need to be well separated.

Although cattle have been shown to have some colour vision, it is not known to which colours they are most responsive (for use on electrified fencing tape and flags for handling).

The eyes of cattle are on the sides of the head. This gives wide panoramic vision of about 300 degrees but mostly with one eye at a time, i.e. monocular vision (Fig. 2.3). This monocular vision means that cattle cannot estimate the size and speed of unfamiliar objects to the sides of their heads. Consequently they can be highly reactive to sudden movement ('spooked'). They only have depth perception at the front where the field of vision of both eyes overlaps to give binocular vision. They will stop and investigate shadows when there is a strong contrast of illumination. This can impede the smooth flow of a herd through a race or between pens. They will also stop and visually investigate unfamiliar things which are not frightening. Cattle are likely to be cautious of a novel object until they have fully investigated it, and this impedes movement.

Social Behaviour

Cattle are social animals, forming attachments and naturally living in groups. When a group is being formed or added to, there is usually conflict between animals or agonistic behaviour. This is behaviour of fight, flight and associated reactions like withdrawal and appeasement. In this way two animals sort out which one has priority of access to something desirable like a patch of grass or a lying place. Cattle are said to have a social order, but it is doubtful if they have this abstract concept. We can construct one from observing the results of a series of agonistic interactions, but cattle probably decide what action to take on the basis of past experience with another beast or one of similar appearance.

When an animal is taken out of the group for more than a day, there is usually considerable agonism on return. This can result in a drop in milk production in cows or physical damage amongst bulls. The animal should, therefore, be returned as soon as possible. If it has been away for more than a day, it is better to return it when the group is intensely engaged in some activity which leaves no spare time to engage the returned animal in a fight or to ride it. Intense grazing periods, such as happen on a new break of grass, dawn or dusk are good times.

Fighting amongst cattle, particularly bulls, is a feature of the mixing of familiar and unfamiliar animals. This is high arousal, which causes much muscle activity and a high incidence of high pH carcases (dark, firm and dry meat).

As cattle are accustomed to being in groups and have a tendency to follow, they are more easily moved and handled as a group. A group will usually move well once one or two have started, and that can be deliberately done when handling. They normally keep at least one other beast within vision and this could explain why curved races are said to work so well for cattle.

There is a certain distance that cattle move away from a predator or a person. If this distance is broached, they will attempt to regain it by moving away. This is called **flight distance** and it can be deliberately used when moving in towards them or moving them on. It varies with experience of handling from at least 100 m for wild cattle, to about 1–7 m for feedlot or milking cattle, and zero for tame cattle, which may, therefore, be very difficult to move.

Cattle normally keep other cattle and people a minimum distance from their heads. The distance is called **personal space**. It is protected by head threats, often a sideways head toss, and varies from 0 to 5 m. It is useful to think of it as a bubble around the head which the animal tries to protect. The effect can be seen when adult bulls are simultaneously moved through a gateway. Because the gate is usually less than 5 m wide, a fight is likely to occur unless they are run through too quickly for a fight to start. Attempts to protect personal space can also be seen when a beast's head is approached when it is tied up. Unless it has been handled frequently and gently it is likely to toss its head around violently.

Experience

Cattle readily learn to accept routine. This is shown by the effects of twice-daily milking and the ease of dairy cow handling. They become so used to routine that small changes in the environment or technique, such as a change in milker, increase arousal and cause a pronounced drop in production. Cattle can also be trained to be led by a halter. This regular handling is not usually possible with beef cattle, particularly those grazed on ranges, some of which may only be handled once a year or even only once in their lives. It is often said that cattle are slow to learn things such as halter leading. However, this is probably a reflection of how unnatural this behaviour is, rather than an inability to learn. In fact, cattle have shown by their aversion that they can remember one bad handling experience for at least three years. Also, once cattle have successfully avoided some undesirable handling, for example being driven into a yard, it is much more difficult to get them back into that area again. So the handling system must be set up correctly at the outset and made to work first time.

Because they are usually handled less, beef

cattle are more easily aroused and frightened by the procedures. It is, therefore, best to restrain them by using confining pens.

Unless treated calmly and gently, even dairy cows which are handled twice a day can be extremely nervous and vigorously resist the routine. Dairy beasts respond well to stroking; once they have come accustomed to it, it calms them down.

Early experience is important. Stock handled quietly from early life are easier to handle later, for example for loading into transport. It is preferable, therefore, for the management of all cattle that they are occasionally handled through the yards.

Handler Characteristics

The attitude of the handler is important. Impatient handling may cause injury (see p.20) and affect production by dairy cows. A confident, introverted milker on average obtains higher yields.

It is often claimed, though it is unproven, that animals detect lack of handler confidence by the smell of human fear. Common observation certainly shows that handlers lacking in confidence do not have good control or get the desired responses easily from animals. However, such handlers are unlikely to be carrying out the techniques as competently as a confident person. The non-confident handler's actions are likely to arouse the animal and cause undesirable responses. A new handler should be taught the proper techniques from the start and keep practising them in a calm environment. The repeated use of low frequency sounds when close to cattle soothes and accustoms them to the handler's presence, as well as having a calming effect on the handler.

Handling with the minimum of disturbance increases the opportunity for heat detection in the dairy herd. Their response to the stimulation of handling may well uncover conditions such as hypomagnesaemia or lameness.

Fortunately, handling can be learned, although for some, who have a natural talent for observing and predicting cattle behaviour, it is easier than others.

General Control

Terms

Different words are used in English-speaking countries for the same handling methods and devices, some of which are not common in all countries. To simplify matters, one term will be used which is believed to most clearly describe the action or device. These terms and their near equivalents or meanings are:

Collecting = mustering, rounding up, bringing in

Confining Pen = crush, squeeze chute, stocks (a pen just wide and long enough to hold one animal with one or more of the following:
- —pole to prevent rearwards movement
- —neck clamp
- —moveable sides to squeeze the animal's body
- —structures to which feet can be tied
- —wide belts to support the animal)

Crowding Pen = forcing/feeding pen (a small pen in which animals are crowded and stimulated to move into a narrow race, usually single-file)

Electric Goad = electric prod, 'hot shot'

Neck Clamp = head gate, head yoke, head bail, stanchion

Nose Holder = nose tongs/lead/grips/pliers

Penning = putting animals in pens either singly or in groups

Race = chute, alley (a long narrow structure which confines animals to moving either single-file or several abreast)

Raised Walkway = catwalk

Single-file Race = confining chute (race only wide enough for one animal at a time)

Sorting = drafting, separating (sorting animals into different classes of stock)

Sorting Race = sorting/separating alley (race

	usually about 3–4 m wide used for sorting)
Tilting Table	= operation table (table to which a standing animal can be tied and then tilted onto one side)
Wide Race	= working alley/chute, feeding pen
Vertical Gap	= mangate, manway (vertical gap in the side of a pen wide enough for an adult person but not an adult cattle beast to get through)
Yarding	= collecting and putting animals into yards

Preparing

Successful handling follows good preparation. By thinking through the job ahead, one can be prepared for all the possible movements and manipulations of the cattle.

—what tasks need to be done?
—what is the minimum amount of handling for those tasks?
—who can best do the handling with maximum safety and least time and disturbance to the animals?
—what is the best place?
—is it safe for the handler and the animal?
—which technique gets the job done quickest and with least disturbance to the animal?
—how can the herd be checked after handling?
—how are the cattle likely to respond? (What is their temperament and experience of handling?)
—are there enough people and pens?
—what happens if the herd breaks back when you try to drive them into the yards?

Handling equipment for cattle must be the appropriate size, strong, well maintained and understood by all who operate it. Dairy cattle are usually tied to a fence, put in the normal restraint for milking, or put in a neck restraint. Beef cattle are usually put in a race or roped.

Moving

When preparing to move cattle there are six questions to consider:

—what is the plan of movement?
—are there enough people?
—is the route clear of obstructions, distractions and projections which would impede the smooth flow?
—is the footing adequate?
—are the cattle movements predictable?
—is the route ready? (for instance, are all the necessary gates open or closed?)

Cattle should be moved at their own pace as they may injure themselves when running. If they do run in the correct direction, let them go.

In moving cattle, patience and silence are a virtue. Impatience by the handler in driving along the track to the milking area has been shown to be a major factor in dairy cow lameness in New Zealand. The signs of impatience are that the cattle swap sides, have their heads up and lean into one another. The other two most important risk factors are poor track maintenance and the presence of a biting dog.

Milking cattle soon learn the routine and become conditioned to collect in the feeding or grazing area ready to move to be milked. The conditioned stimuli for this are those associated with the routine such as the approach and noises of the collecting person.

It is best to move cattle in one bunch rather than repeat the process with several smaller groups. A group has a collective flight distance and also a point-of-balance. It can be moved by applying stimulation ('pressure') to the collective shoulder of the group to keep those at the front moving in the required direction. Constant vigilance is necessary for changes in movement of the bunch or breakaways by individuals.

For movement through gates, the lead animals should be moved through and then others allowed to follow at their own pace, driving them on only if the flow stops. Too much stimulation at this stage is likely to cause breakaways. Once a group is moving, try and keep them together at a speed comfortable for them.

The amount of movement should be kept to a minimum. The faster animals move, the more corners they have to turn and the narrower the gaps, the greater the chances of bruising. The better the yards are designed, the lower the stimulation needed to move them. All agents of movement need to be well controlled. Poorly-trained dogs can cause a bunch to scatter. Cattle can be trained to follow feed by conditioning them to a call when feeding and when they are hungry. To start the training, make the call, rattle the bucket, drop a little food on the ground, move a few paces in the required direction and repeat the process until they start to follow.

Driving and droving

Local regulations for the moving of stock along public roads should be checked. It is necessary to have at least two people in control when driving cattle along a public road in the United Kingdom. One is needed at the front to warn oncoming traffic and the other should be behind to drive the cattle. Cattle must not be allowed to do damage or trespass.

Aids

Care is needed with sticks as they readily bruise animals. They can be used to move animals with a light tap on the rump but canvas flappers (Fig. 2.4) are just as effective and do not cause bruises. Electric goads are considered to be excessive force and they cause cattle to run and dart. They are not only inhumane but they are often counterproductive. Although they may be useful for overcoming reluctance to enter a race, a canvas flapper, rolled up newspaper or house broom is also effective.

Individuals can be moved forward by twisting the tail with a loop or flattened 'S'-shape (Fig. 2.5). Great care should be taken to hold the tail about 15 cm from the base and to bend it gently as the tail can be broken. It is safer to stand close beside the tail and not behind.

In many circumstances, dogs are best tied up or confined when cattle are being moved in buildings or yards. They should only be used if well trained and controlled and if the cattle are accustomed to

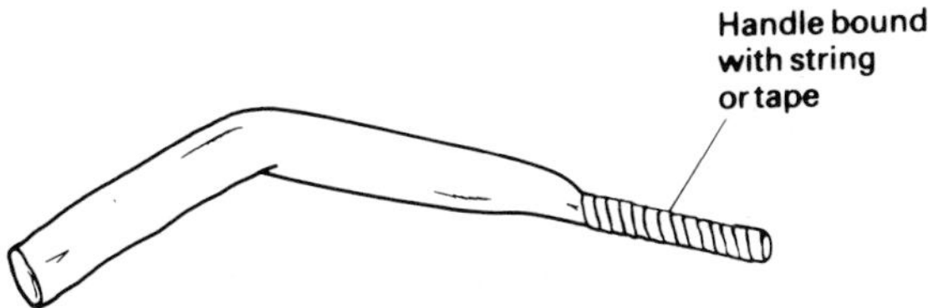

FIG. 2.4 Canvas flapper (made from 7.5 cm firehose).

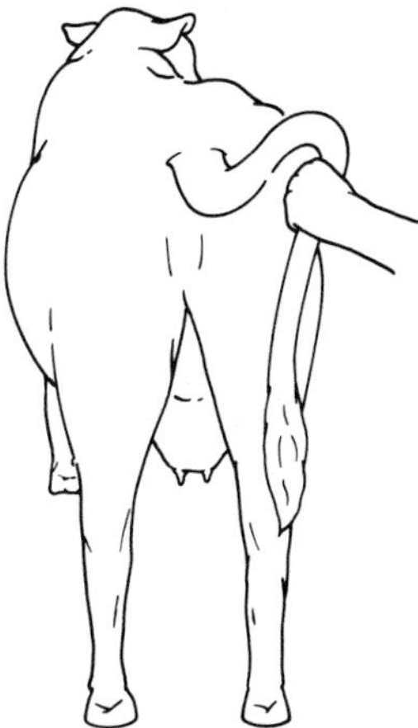

FIG. 2.5 Twisting the tail to make a cow move forward.

dogs. If dogs are needed, it means that the yard design could be improved.

Observation suggests curved single-file races fill faster, are quicker, and are more efficient than straight races. The cattle are said to fill the race up by following the first in who walks round to the end and there is no backing out.

Yard Design

Yards should preferably be built on level ground. Movement should be uphill rather than downhill. Circular yards are believed to be better than rectangular. Using curved races means a person on a raised walkway can easily keep in the optimum position of about 30 degrees from the midline when following behind a beast.

Sliding gates are effective for closing off single-file races, e.g. between the race and the confining pen or weighing area. They can be opened or shut from outside the race and, when part closed, are less likely to be forced open again by a backing beast. Vertical gaps have a dual function: they allow very easy movement between areas and they can be a life-saving escape route (Fig. 2.6). If there is a choice, it is probably better to have cattle moving downwind.

Sorting

Ideally the separation of cattle into different classes is done in a sorting race, but a wide race or two pens may also be used.

Bruising is highly likely during sorting because cattle are being aroused and moved through narrow gaps with sudden changes of direction, so the conditions listed previously for moving cattle are all applicable. The following, however, have specific relevance:

—gates should open outwards from the sorting race, with the person opening it on the outside
—groups should be of a manageable size
—drafting should be back towards the entry point
—use should be made of flags (Fig. 2.7), canvas flappers, or sticks to extend the range of control.

The handler's position in relation to the animal's shoulder can affect which way it will go. An imaginary line across the animal running through the shoulders can be considered its point-of-balance. Moving behind the line causes the beast to move forward and moving in front of the line causes it to go backwards (Fig. 2.8). Standing on this line results in neither forward nor backward movement. From the front, cattle can also be deflected sideways by moving either side of an imaginary line drawn through the middle of the animal's length. This cannot be done from directly behind, as the handler will be in the blind spot (Fig. 2.9).

Sticks and pipes increase a person's profile and thus control of cattle movement is increased by extending two pipes fully to either side as shown in Fig. 2.10. Conversely, this effect is reduced by lowering the pipes and turning sideways onto the cattle as shown in Fig. 2.11.

As a general principle, it is better to remove the quieter animals from the more excitable — cows from bulls, cows from calves, and old from young. The quieter class of stock is thus easier to handle and the more excitable are disturbed as little as possible. Cows should be removed from their calves as cows move more easily on their own than with calves.

There is a tendency for moving cattle to follow one another and, if necessary, decisive, firm handling is needed to stop this. A flag dropped in front of an animal's head is an effective way to stop it and send it back (Fig. 2.7).

Approaching Individuals

It is essential to talk quietly, preferably with low-frequency sounds, when approaching an animal. This avoids surprising it with the consequent risk of it kicking out or running off. Regularly handled cattle should be approached in the way and from the side to which they are accustomed. Traditionally this is from the left but particular circumstances, like a milking system, may require

Fig. 2.6 Handler escaping through a vertical gap.

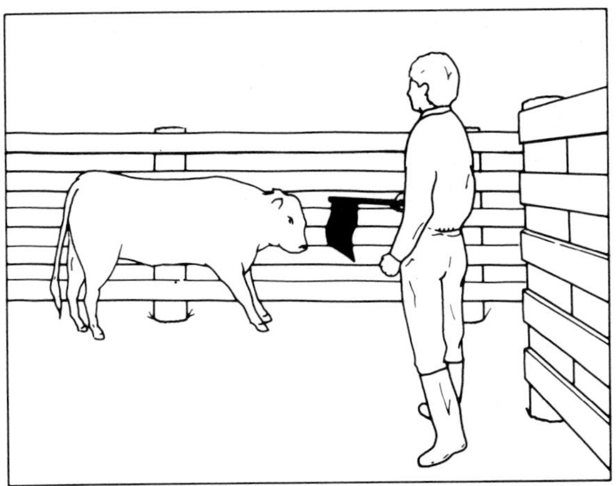

FIG. 2.7 Using a flag to stop a cattle beast going through a gate in the pen.

another direction. Touch the shoulder area first and gently move up to the head. Fearful, flighty or aggressive animals are likely to move or kick sideways. The kick is usually a semicircular slicing action which can cause serious injury. Aggressive animals like dairy bulls and nymphomaniac cows are more likely to butt and squash a person. An escape route should be close by before approaching such animals. A direct approach is aggressive to cattle and this may be met by aggression, i.e. a charge forward and butt.

Restraining the Head

The degree of restraint needed depends on the tameness of the animal and the disturbance caused by the manipulation. Some or all parts of the body may have to be restrained for the safety of both animal and handler.

The head is the first and most important area to restrain as it can seriously injure the handler when it is violently tossed about. Having restrained the head, the animal is much more likely to be still. If the head is not restrained, many cattle continue struggling even though other parts of the body are closely restrained. It appears as though they think they can break free by struggling when the head is free. Heads of cattle may be restrained by:

—putting into the usual milking area
—a self-locking neck clamp in the feeding passage
—haltering
—tying the neck to a fixture such as a stanchion or neck chain
—clamping the neck in a yoke or stanchion and

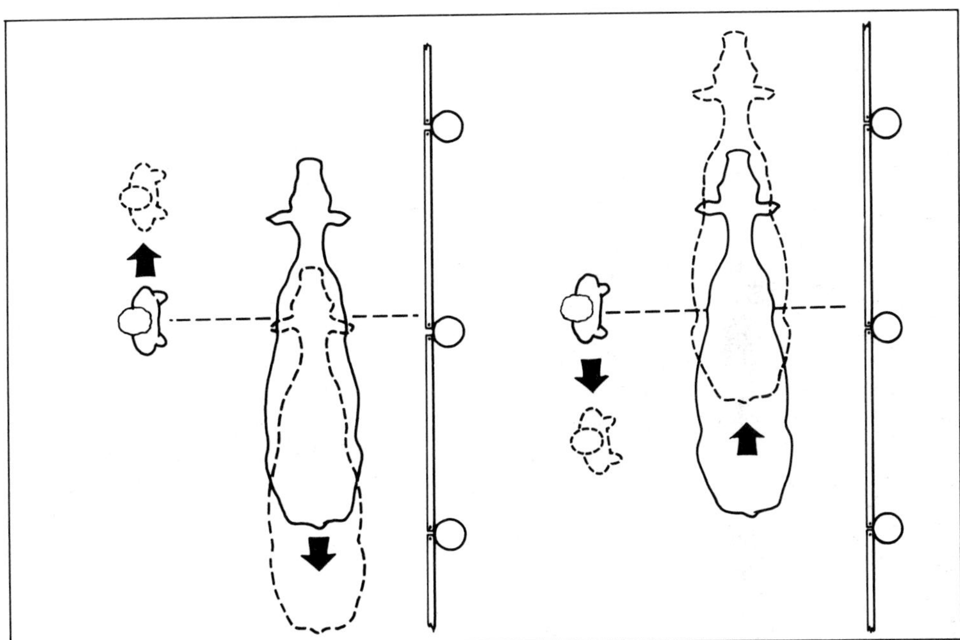

FIG. 2.8 Using the point-of-balance through the shoulder to move a cattle beast either forwards or backwards against a pen side.

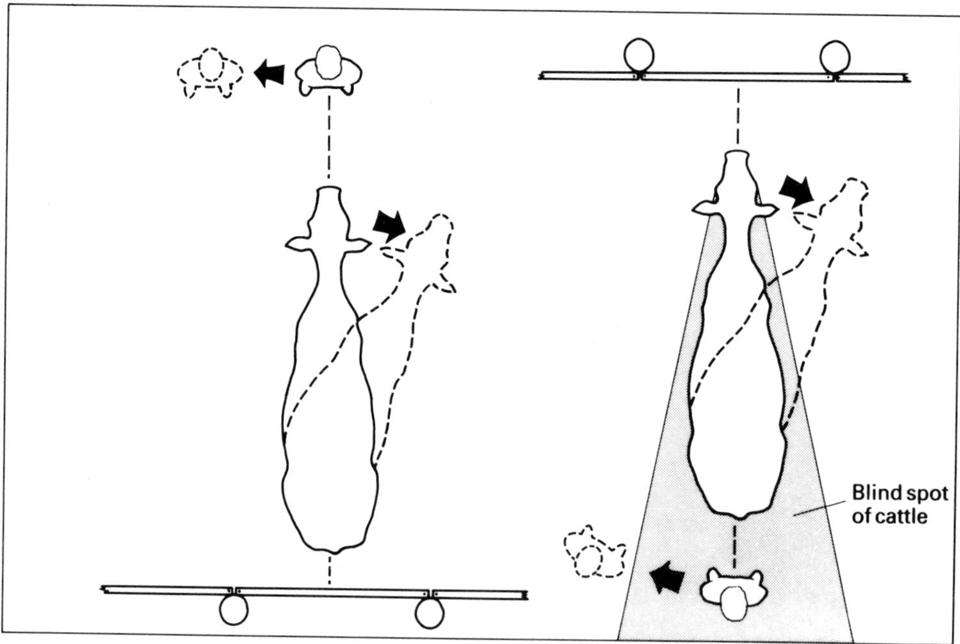

Fig. 2.9 Using the point-of-balance through the mid-line to move cattle to the side from either in front or behind. Note that cattle cannot see the handler standing in the blind spot directly behind them.

either holding the head down with a nose bar (Fig. 2.12), tying it up with a halter, or putting pressure on the nasal septum.

Haltering

The halter can be an effective, cheap, and humane restraint, particularly for animals trained to it. With infrequently or poorly handled animals,

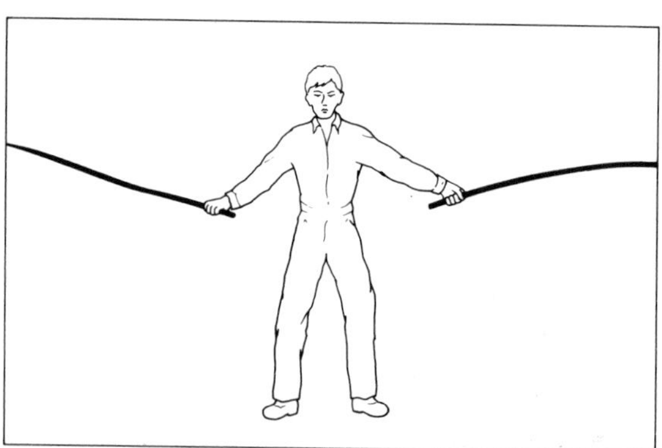

Fig. 2.10 Increasing the range and power of the handler by using two sticks.

Fig. 2.11 Reducing the body profile of a handler to reduce power of movement. Note that the side is presented and the stick lowered.

however, putting the halter on may be a difficult and lengthy procedure. The halter should be made ready with an enlarged head piece and nose loop (Fig. 2.13). Haltering unhandled animals is best done either by confining several in a small space or, individually, in a confining pen, dropping the enlarged nose loop over the nose and pulling the head piece over the poll (Fig. 2.14). Alternatively, the beast is caught by a loop round the neck and, once calmed down, it can be haltered. If the halter is to be left on, a half hitch should be tied where the rope leaves the halter to become the lead rope and a wisp of hay may be introduced to ease the release of the knot (Fig. 2.15).

As a general rule, the animal should be approached slowly from behind the shoulder at 45 degrees, which is a relatively safe position. Talk quietly to the animal, scratch its back and keep moving with it.

Some animals keep dropping their heads when the chin loop is dropped in front of them. For them the halter is slipped over the ears first and, when they lift their heads, the chin loop drops under the head and the lead rope is tightened. The key things to haltering are confinement, patience, movement with the animal, and leaving the halter part on with a slack lead rope until the last moment.

Nose

Pressure on the nasal septum is a very effective way of controlling head movement. A large animal can be gripped in that way for a few seconds using the thumb and forefinger. But for longer periods and greater security with large animals, a nose holder is needed (Fig. 2.16). This device is available in a variety of forms. The best nose holders have the following features:

—rounded smooth bulbous ends that go into the nose (so that the cartilage is not damaged or destroyed)
—a gap of about 3 mm between the bulbous ends
—a device for holding the handles in the closed position so that they do not have to be held.

Some devices have long handles to give a mechanical advantage to the handler. Others have short handles and ropes so they can be wrapped once around a fixed structure (Fig. 2.16). The end of the rope should be held so it can be let go instantly. It should never be left unattended. If the beast goes down, the nasal septum and fleshy exterior part in the middle of the nose may be ripped. Stand beside the head and facing the same direction as the beast. Put your thumb and first finger into the

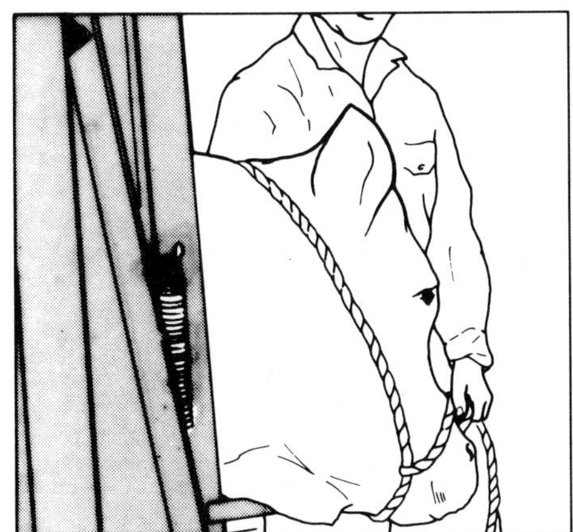

FIG. 2.12 (*left*) Neck and nose bar holding a cow's head in a neck clamp. (*right*) Head restrained with neck clamp and halter.

FIG. 2.13 Halter showing enlarged head and nose loop.

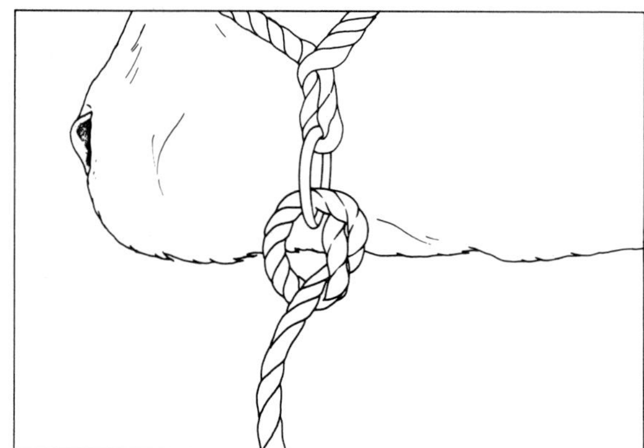

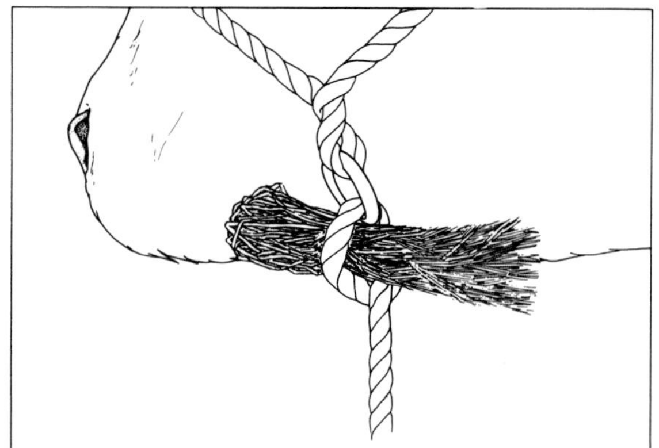

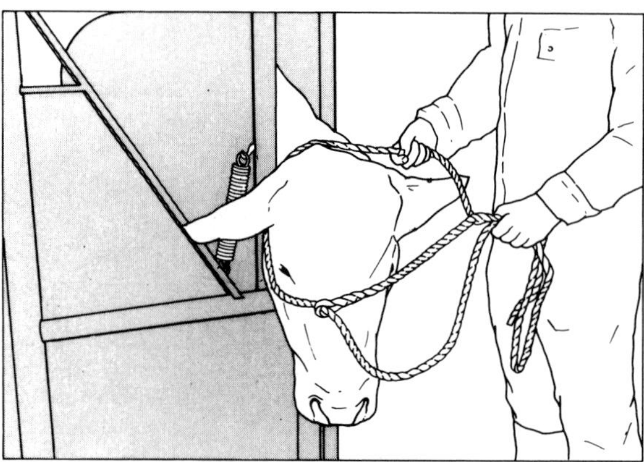

FIG. 2.15 (*top*) Half-hitch on halter to prevent over-tightening or slipping off. (*centre*) A wisp of hay in the half-hitch may ease release if the knot becomes wet or tight. (*bottom*) Slipping a halter over the ears.

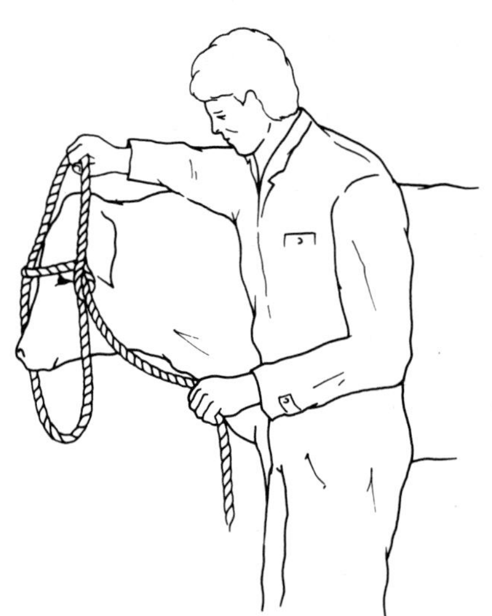

FIG. 2.14 Haltering. Note that the large loop has been dropped over the nose and the ear piece can then be pulled over the poll.

nostrils to restrain the animal whilst the holders are inserted. A bulbous end is inserted on an angle into one nostril and the holder rotated to get the other end into the other nostril past the fleshy exterior. Control is best when the head is held high.

In many countries, bulls from 12 months old must have a nose ring and possibly a chain or rope attached. The ring must be fitted well forward in the nasal septum in front of the cartilage. A pole with a clip on the end can be attached to the nose ring of a bull. This provides a greater mechanical advantage on the nose and keeps the bull's head away from the handler. Ideally the clip can be opened from the other end of the pole so the handler can keep well away from the bull's head.

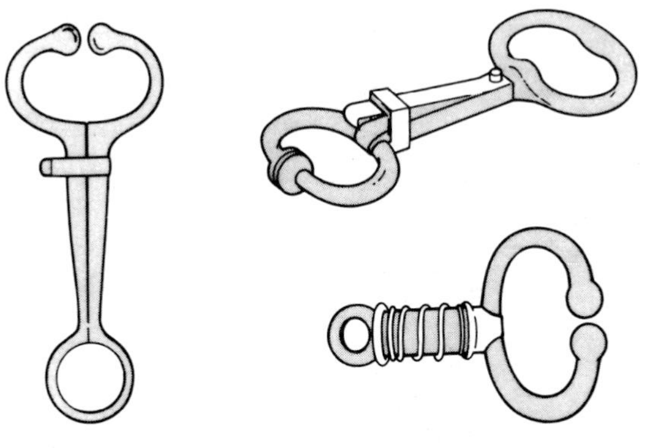

FIG. 2.16 (*top*) Nose holders. (*centre*) Nose holder with long handles. (*bottom*) Nose holder with clip and rope.

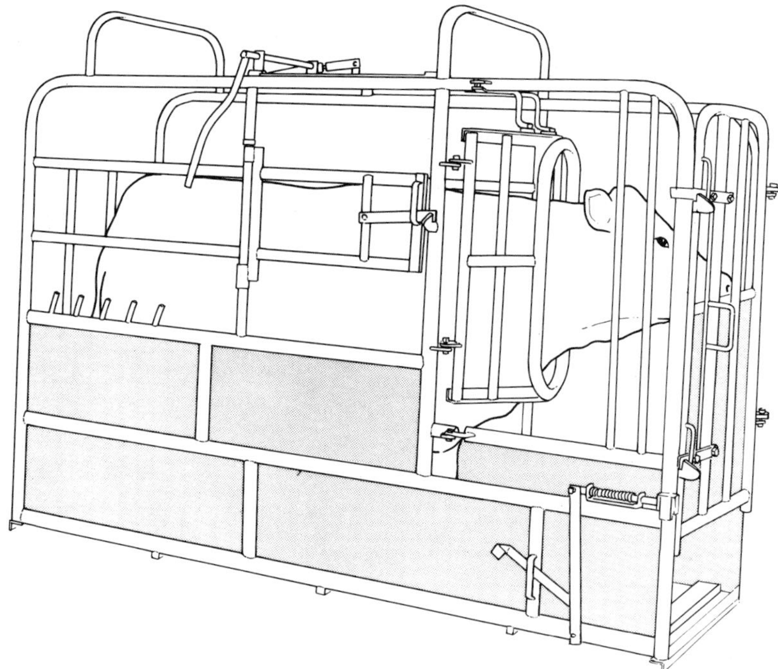

Fig. 2.17 Cattle confining pen of conventional design.

Restraining the Whole Animal

Confining pen

Whenever cattle are to be frequently handled it is a good investment to have a confining pen (Fig. 2.17). This restrains the whole animal and most body parts. A good design has:

—its location at the end of a single file race
—a non-slip floor
—enough space and grip on the floor for cattle to stand up should they fall
—no inside projections
—water supply and drainage
—the capacity to alter the width to fit snugly different sizes of cattle
—a clear space of 6 m visible through the exit gate
—a neck clamp which is quick-release when the job is finished or if the beast becomes distressed
—a vertical closing neck clamp so that a beast going down does not fall onto a bar, its neck is not twisted nor is the head held in a fixed position
—an arrangement where the head does not have to be released back into the crush before the front gate is opened
—an arrangement whereby the head and neck are released before the exit gate is opened as the beast may lunge forward and choke or break its neck
—a neck fastening with bars to hold the nose and the head down so that the head is immobilized
—a rear gate or pole to prevent backwards movement when the neck is not clamped
—secure fixation to the ground so it stays in position under the forces of moving and stopping beasts
—removable side bars and panels to give access to all body parts for surgery (these must be shut until the animal is restrained because feet or legs may get caught and broken in them)
—provision for wide belts to stop the beast lying or falling down
—moveable arms on the front vertical posts for swinging into position for tying forelegs
—a moveable transverse pole behind the animal to hold a leg up in position.

The sequence of activities is: entry, close neck clamp, squeeze in the sides, apply head and then nose bars. To release the animal the procedure is reversed.

Tilting table

Tilting tables offer complete restraint and maximum handler convenience. Cattle are probably comfortable in this position because they are likely to be in a hypnotic state (tonic immobility). Restraint and inversion is a common way of inducing tonic immobility in many species.

The tables may be mobile or fixed. The technique requires head restraint, restraining girths, winches, padding, ropes and rings. The animal is walked on to a base plate with the table in a vertical position. The head is closely tied to the table by a neck clamp or halter. Two restraining girths are placed around the chest and the third between the hind legs and either the udder or the scrotum. The animal should be hooded to keep it calm and to shield the upper eye from bright light. Special care is needed at this stage because cattle are likely to resist being strapped against the table.

The whole structure must be firmly anchored to cope with the weight transfer as the beast is tilted over. The table is rotated into the horizontal position and padding placed under the shoulder and upper foreleg to prevent damage to the radial nerve. The feet are then tied as required.

The reverse procedure is used to get the animal back on its feet. Cattle should not be held in lateral recumbency for more than 30 min because of the build-up of gases in the rumen.

Cattle can be forced to lie down and be immobile by using a rope to put pressure around the thorax and abdomen. It is not now commonly used because of the effort involved, the risk of damage to the animal, greater use of confining pens, and the availability of chemicals for casting. The physiological mechanism is not understood. A common roping pattern is shown in Fig. 2.18.

If it is necessary to cast with a rope the following should be considered:

—a person should be put in charge of the head to control where the animal falls, that person stands to the right side because the beast should be laid on its left side to reduce abomasal displacement, the head and neck are turned backwards and the person leans into the shoulder (Fig. 2.18, centre)

—if there is nobody to hold the head, it should be tied low to the ground

—the area should be firm but not hard, and be flat and free of injurious material.

A loose bowline (Fig. 2.18) is put around the neck and two half hitches around the body, the first behind the shoulder and the second just in front of the tuber coxae. The slack is taken up from the hitches and the free end is pulled steadily backwards until the beast lies down. If it resists going down by straddling its fore and hind limbs, they can be roped close together (hobbled). Once down, the upper legs should quickly be restrained to reduce the chances of the operator getting kicked. The person in charge of the head can kneel on the neck and hold the upper foreleg flexed. Occasionally the hind leg may strike out and, if necessary, fore and hind legs can be tied together.

Large bulls need considerable strength and patience. A small beast can be cast using the lark's head twitch around the thorax and abdomen as shown in Fig. 2.19. Pressure is applied by pulling up on the rope.

Other methods of casting cattle have been discussed by Fowler (1978), Leahy and Barrow (1953), Miller and Robertson (1959), and Stöber (1979).

Casting is not recommended for pregnant cattle because of the possibility of abortion. It can also result in displacement of the abomasum, bloat and pneumonia. A careful watch should be kept for respiratory distress, loss of consciousness or bloat. In any such event the beast should be allowed to rise and recover.

Preventing Kicking

The dangers of cattle kicking were mentioned in the introduction. The only guaranteed method of stopping kicking is to tie up the legs or chemically immobilize the animal. There are, however, several methods of reducing the chances of kicking.

Tail lift

A simple technique for preventing kicking is to lift up the tail as shown in Fig. 2.20. Where cattle cannot move forwards or sideways this is a very efficient restraint. Great care must be taken not to damage the tail. Stand close to and directly beside the tail. Use one hand about 15 cm from the base to

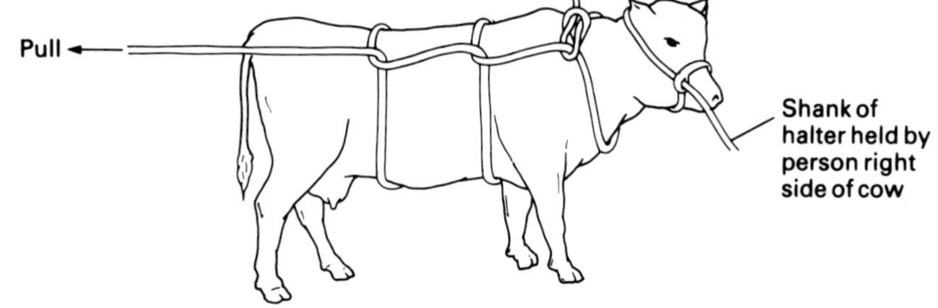

FIG. 2.18 (*top*) Half hitches used for casting cattle (Reuff's method). (*centre*) Handlers in position for casting. (*bottom*) Bowline (detail).

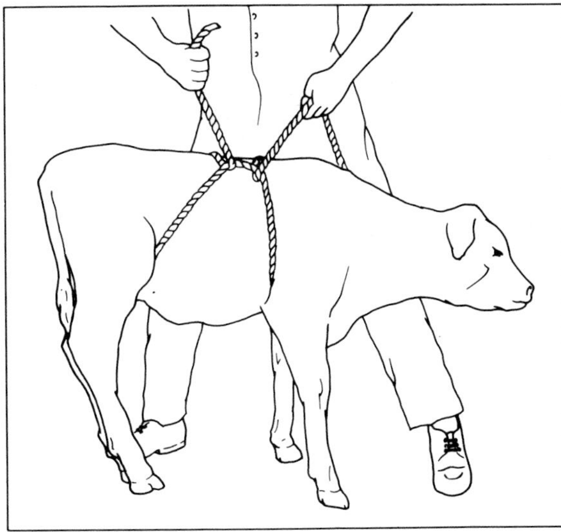

FIG. 2.19 Lark's head hitch used for casting a calf.

lift the tail straight up in line with the backbone. The other hand may be used for support. As with other restraints it should only be used as necessary and eased off when the animal stops kicking or moving about. Cattle will generally not kick when this is applied.

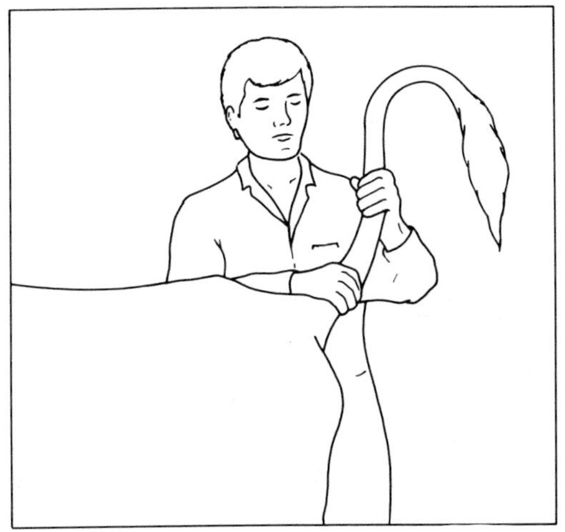

FIG. 2.20 Tail lift. Note that the upper hand is positioned about 15 cm from the tail base.

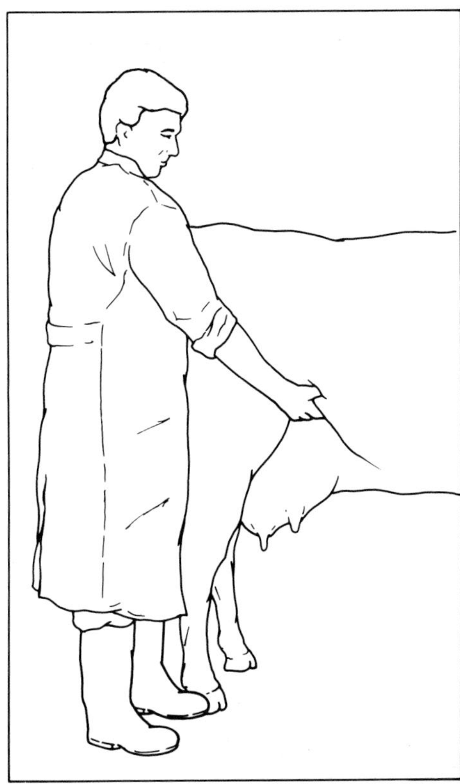

FIG. 2.21 Grasping and lifting flank by a fold.

Flank pressure

Pressure on the flank also reduces the chances of kicking. This can be done by lifting the flank fold up by hand (Fig. 2.21) or by a C-shaped clamp placed in the flank area and over the opposite loin (Fig. 2.22). A rope may be tied around the abdo-

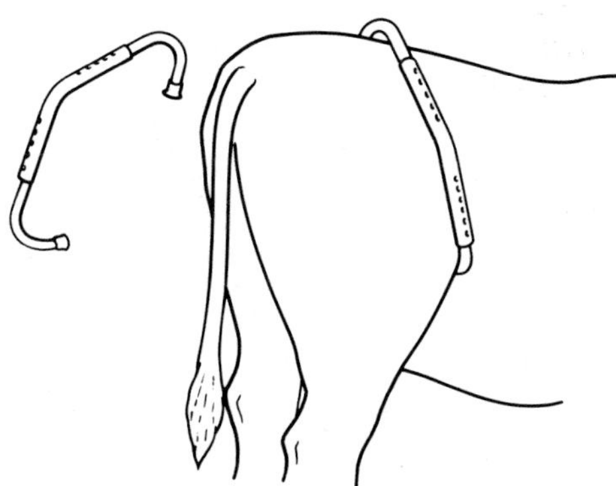

FIG. 2.22 C-shaped clamp (immobilizer) for preventing kicking.

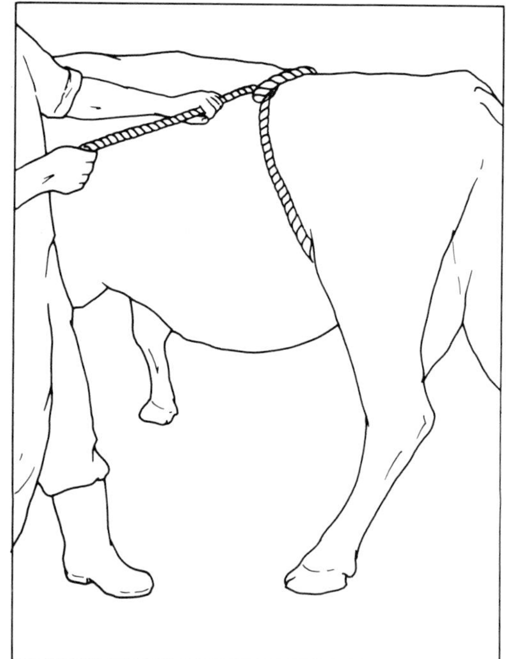

FIG. 2.23 Flank pressure exerted by a rope around the abdomen.

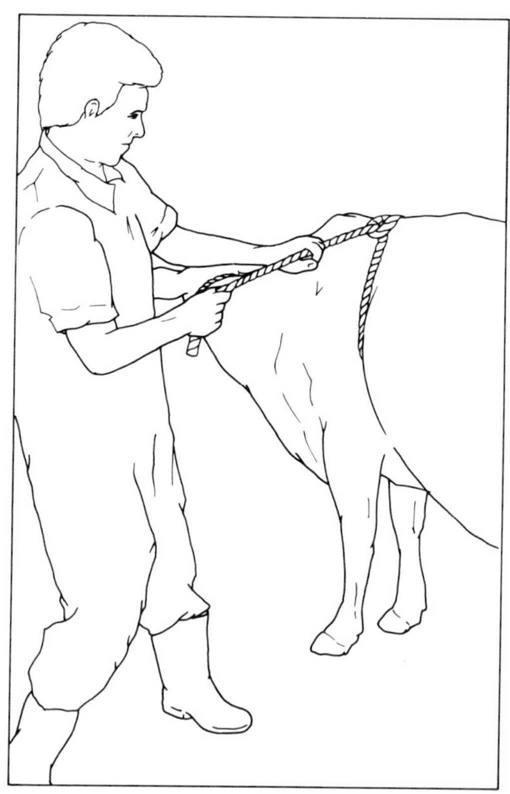

FIG. 2.24 Pressure applied over the back and chest by a rope.

men, as shown in Fig. 2.23, but there is a risk of damage to the milk veins or penis.

Restraining legs

The hind legs can be tied together (hobbled) but the technique should only be used in emergency. The handler is in danger whilst putting them on a kicking beast. Cattle can become highly agitated during or after application and so constant supervision is essential. The cow and the handler may be injured if she falls over. A number of other devices have been developed for preventing kicking at milking. Clamps and U-shaped devices can be put on the Achilles tendon just above the hocks.

Ways of restraining legs will be described later under **Common Manipulations, Feet.**

Chest twitch

A rope looped around the chest behind the forelegs and pulled tight (Fig. 2.24) may calm cattle which have not settled down when the head, limbs or tail have been held.

Tying Tail

A tail lashing about is dangerous and it should be held or tied out of the way. If it is tied, it should be to the beast's own body. Never tie it to a fixed object as the tail may break if the beast falls down. A sheet bend can be made with the end of the tail and a rope, which is looped and tied around the animal's neck (Fig. 2.25).

Common Manipulations

Whole Animal

Weighing

Animals other than small calves, are usually weighed by scales in a weigh pen positioned in or at the end of a race. The weigh pen should have good visibility through the exit gate so the beast enters readily. For accuracy it is essential to zero the scales at the start and frequently thereafter. The platform may rapidly accumulate mud and faeces,

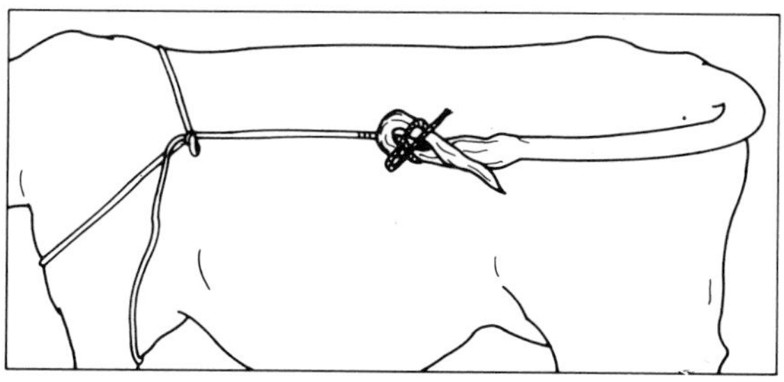

Fig. 2.25 (*left*) Tail tied to a rope around the neck. (*right*) Sheet bend for tying rope to tail.

in which case cleaning and zeroing may have to be done for every animal.

Some cattle will not stand still in the pen which makes accurate reading impossible. Scratching their backs or placing a hand near the head often settles them down long enough for a reading.

Raising an animal that has gone down

Making it stand. Before attempting to get an animal to stand, it is vital to examine it to check that it is humane for it to do so and that standing is possible. It may be down because of insufficient blood supply, paralysis, fracture, dislocation or muscle damage of the legs and an attempt to get it to stand is a diagnostic test to understand why it is lying.

To be able to stand, cattle need:

—functional limbs; after lying for some time the limbs under the body may be stiff or numb, so the animal should be rolled on to the other side to relieve pressure on the underneath limbs
—limbs flexed underneath it, so put the legs in the natural position for rising
—non-slip footing to provide purchase for the feet
—sufficient room to rise.

Various stimuli have been successful in getting cattle to stand. Examples are:

—clapping and shouting close to an ear
—slapping the neck and chest
—close proximity of a barking dog
—pouring cold water into an ear
—pressure on the tail on a hard surface.

During handling in a confined space like a confining pen, cattle will sometimes go down. They may lie there for many minutes and refuse to get up despite the usual stimuli. A technique which is frequently successful is to block the flow of air through the nostrils. The palm is placed on one nostril and the fingers used to close off the other. The other arm is used to hold the head which is likely to be tossed around as the beast tries to breathe. If it tries to breath through the mouth, try holding the mouth shut.

Help should be given to steady the cow as she rises. Priority should be given to holding the tail near the base. It is the best manual support that can be given and the tail is less likely to be broken when held near the base. Other assistance can be directed to support lateral movement as she rises. The tail should not be used for lifting the cow up as that is likely to cause damage.

Lifting. Recumbent cattle can be lifted by hand, rope, liftbag, harness or hoist. The tail should not be used to lift as there is a danger of causing paralysis of parts in the hind region. Once the animal is standing, the tail should be held to steady the beast. Help in standing can be given by lifting with a flank fold grip on both sides. Ropes under the chest and inguinal areas can be used to assist standing and steadying. Inflating a large airbag under the chest and abdomen is a good method. The cow can be lifted in a webbing harness and slung for up to one hour to allow sucking and leg movements (Fig. 2.26). It can also be lifted by a clamp over the pin bones (tuber coxae) (Fig. 2.27) but this should only be used for a few minutes because of the possibility of suffocation, circulatory strain, and damage to the muscles and skin over that part of the hips. Lifting attempts should be used cautiously. They will harm the beast when done repeatedly, unnecessarily or incorrectly.

Once standing, the animal may need assistance for some minutes until it is competent at walking again. Help can be given by holding the base of the tail or lifting the flank. It is essential to have a good ground surface so that the beast does not slip. The environment should be safe and comfortable for the first few days afterwards.

Moving a 'downer'

Cattle can be rolled onto heavyweight mats or skids which are then dragged along the ground. As long as the beast is comfortable during the process, the greatest danger is probably to people who injure themselves by dragging with a bent back. It may be better to use a vehicle to pull the structure.

Head

Drenching

Before doing any drenching, read the instructions. Some preparations should not be mixed and some have special precautions.

It is worthwhile being patient and gentle when drenching, particularly when it is to be done frequently. Cattle can become very difficult to handle once they have developed an aversion and have learned how to avoid drenching. Drenching must be done with the animal standing so that it can swallow normally and not get liquid into its lungs. The head needs to be still.

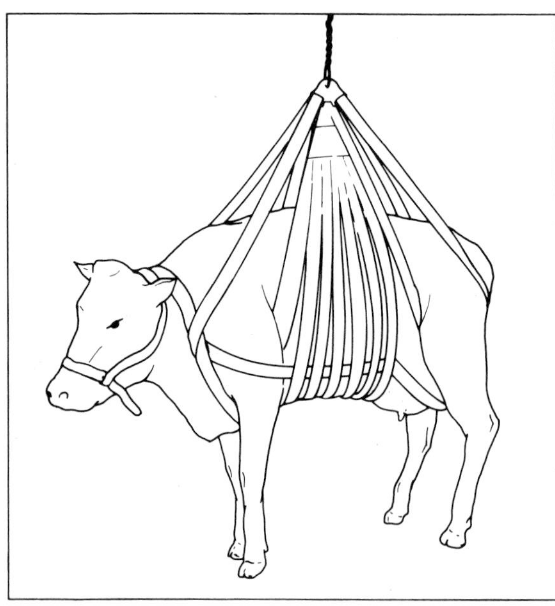

FIG. 2.26 The 'Downacow' harness for lifting recumbent cattle.

Calves can be drenched in a race when packed tightly enough for a person to be able to restrain one but with enough room for the person to move. It is best to move back through the group and wear leggings as calves scrabble and kick with their hind legs.

Heavier cattle need restraining by the neck — the method depends upon how used they are to drenching. Dairy cattle can be dosed in the milking area but beef cattle will probably need a confining pen with a neck clamp. The handler stands alongside the head, facing the same way (Fig. 2.28) and grasps the head with the inside arm cupping the jaw. It is risky and unnecessary to put your fingers in the beast's mouth. The other danger to the handler is sudden, violent head movements. To reduce the movements the beast should be moved back as far as possible in the neck clamp. Putting your inside leg under the head, insert the drenching device (gun or bottle) into the mouth between the molars and incisor teeth, taking care not to knock the teeth. Put the end over the middle of the back of the tongue and deliver the drench slowly so it can be swallowed. If it goes too fast, it may go into the lungs causing coughing and pneumonia. Remove the device taking care not to knock the teeth. Pause a few seconds to check that all the drench has been swallowed.

If you are using a bottle, take particular care not to get the neck between the molars where the glass can be broken; it is safer to tape the neck of the bottle or fit a rubber tube (e.g. a teat cup liner) to minimize damage or breakage.

Drenching with paste, bolusing and giving capsules are done in the same way. The item to be swallowed must be small enough and the delivery system gentle enough so the throat is undamaged. It is important to release the item slowly and gently into the back of the tongue so that there is no damage and the beast can swallow it. Watch to see the item is swallowed. If it has not been ejected in a few seconds, one can assume it has been swallowed. To open the mouth with a wide bore pipe for bolusing, it is necessary to put pressure on the

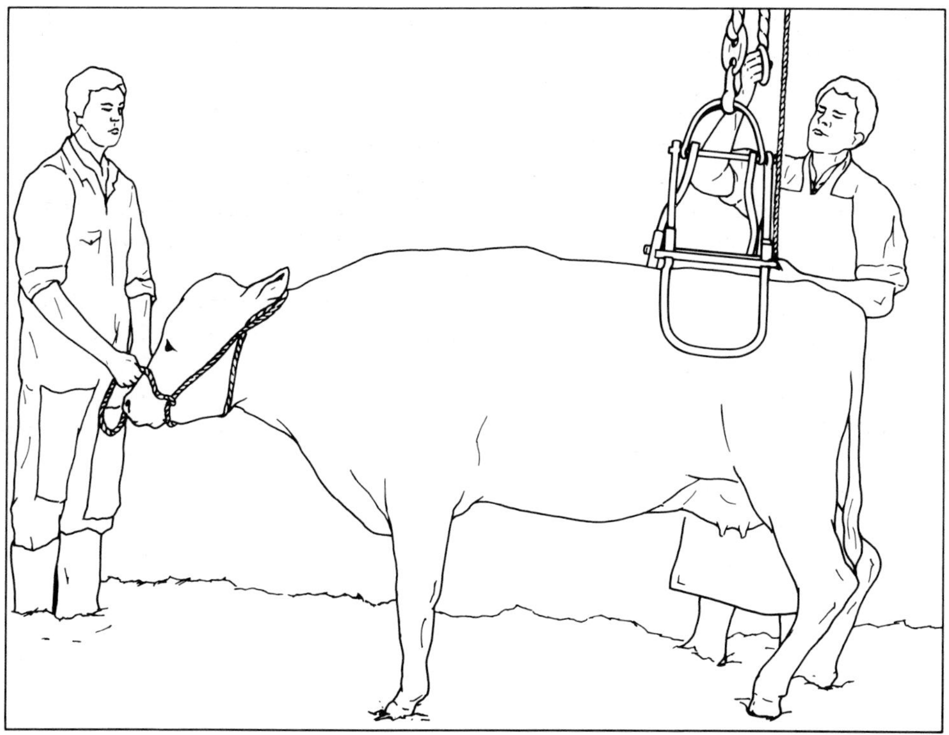

Fig. 2.27 The Bagshawe hoist used for raising cows with posterior paresis.

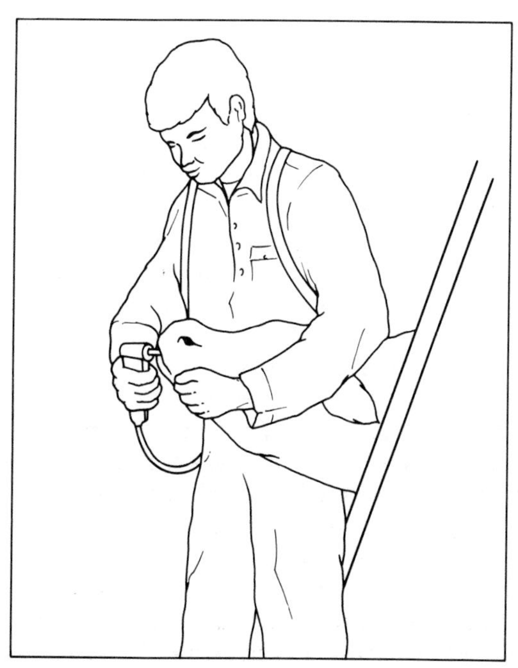

Fig. 2.28 Restraining the head for drenching, using a neck clamp. Note that the fingers and thumb are outside the mouth.

gums in the dental space — the gap between the incisor and molar teeth — keeping the fingers and thumb well clear of the teeth.

Examining mouth and pharynx

A dental wedge is used for examining the mouth and pharynx. To insert the device the head should first be held between your body and arm. The arm is placed over the head and the lower jaw gripped so the head is pulled against your body. The wedge is placed between the molars on one side to keep the jaws apart to allow examination (Fig. 2.29).

It is dangerous and unnecessary to put your fingers in a beast's mouth to open it. The mouth can be opened by inserting the wedge between the gums where there are no teeth (dental space between the incisors and the molars).

Stomach tubing

Getting a tube into the stomach requires good head restraint. More than a halter is required. Up

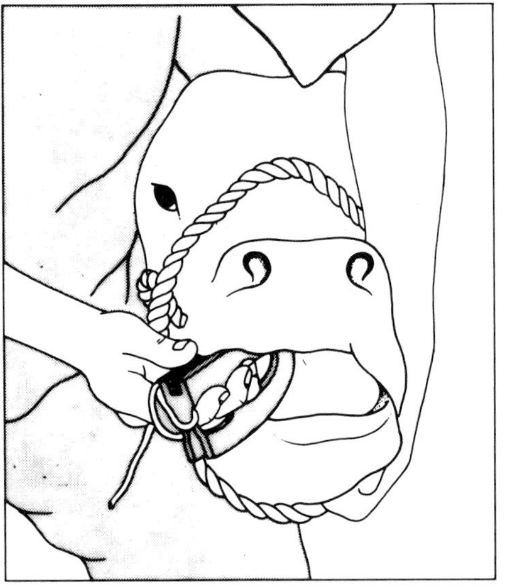

FIG. 2.29 (*left*) Inserting a dental wedge into the mouth to fit between the upper and lower molars. (*right*) Dental wedge (Drinkwater gag).

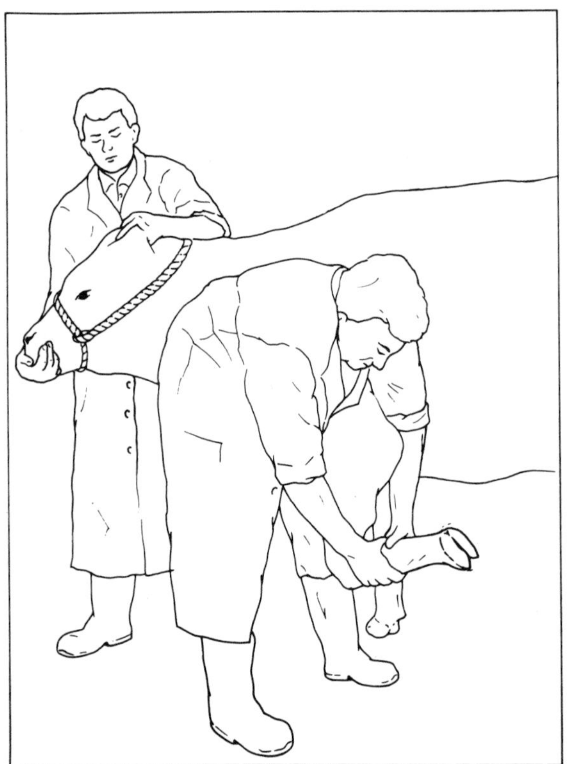

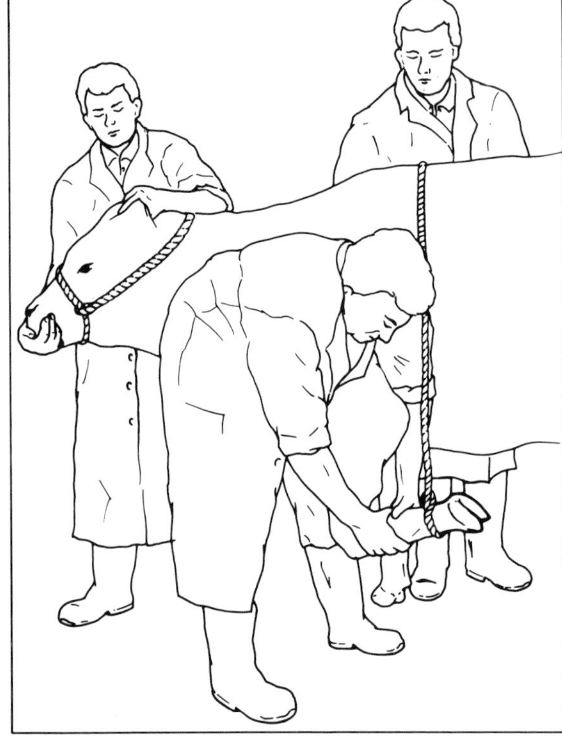

FIG. 2.30 (*left*) Raising a foreleg by hand. (*right*) Raising a foreleg with the help of a rope over the withers.

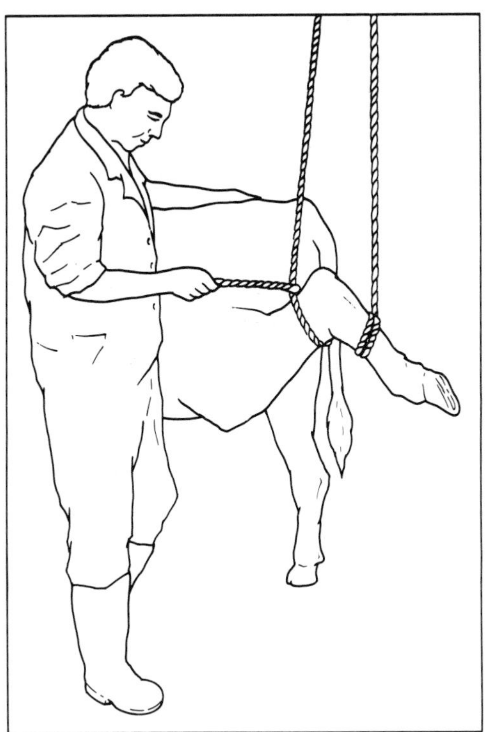

FIG. 2.31 Raising a hind limb using a rope over a ring, hook or beam in the ceiling.

to about 90 kg (200 lb) may be held by a person standing astride it pulling it back against a solid structure like a fence or wall. Heavier animals need to be dosed with the head fixed in a neck crush with head and neck bars, or an equivalent restraint.

The end of the tube should be rounded both on the internal and external edges. Sharp edges may damage the delicate lining of the pharynx, oesophagus and reticulo-rumen.

The mouth should be wedged open as described above. A tube is fed into the mouth along the top of the tongue and down the oesophagus. Its presence in the oesophagus can be felt through the skin of the neck. If it is in the trachea, it cannot be felt but air will be passing in and out.

To administer a fluid the head should be tilted slightly upwards. Liquid is poured down the tube as fast as it can go from a funnel.

Feet

Lifting

There are several methods for lifting a foot off the ground. One reasonably strong person can lift and support a forefoot for a few minutes (Fig. 2.30, left). If you give more than very light support, the beast may put its normal weight on that limb. A straw bale can be placed under the knee. Forefeet can be lifted with a rope over a beam or over the withers (Fig. 2.30, right). A rope with an eye-splice at the end is preferable. A loop is made around the metacarpus. The free end of the rope is looped over a strong fixture, preferably directly above the leg, and the foot pulled up by tension on the rope. The free end should be held by a person so that it can be released instantly.

A hind foot is usually lifted by rope. A noose is formed around the metatarsus. The free end is looped once around a fixture, which should be above and behind the leg. By pulling on the free end, the leg is lifted up. It is held in position by a helper who wraps the rope around the Achilles tendon and holds the free end (Fig. 2.31). The rope can be let go quickly to reduce the risk of injury should a beast fall.

Two strong people with a stout pole may be able to lift and hold a hind leg satisfactorily. They should put the pole horizontally in front of the hock and lift the pole up and back. They should support the beast by leaning in towards it. A twitch can be tied around the Achilles tendon to reduce kicking with that hindfoot. The footpole can also be used on forelegs.

No matter which method of lifting is used in fore or hind legs, the foot should be lowered occasionally to give the animal a rest from balancing on three legs. Manual lifting of cattle feet is only suitable for quiet animals and handlers need to be reasonably strong.

Trimming

Hoof-trimming requires the foot to be fixed. A hoof-trimming pen is a good investment with larger dairy herds where animals are trained to lead. Belts are used to give very slight support without lifting the beast off the ground. The boards at the sides should be rapidly removable with wing nuts and bolts so that a leg can be quickly released. A leg may be broken when caught between the floor and horizontal supports.

Udder

Early handling

Handling of the udder is made much safer and easier if cattle are accustomed to it from an early age.

Treating mastitis

Mild mastitis can usually be treated with minimal restraint, usually just standing in the normal position for taking a milk sample. However, as the udder may be painful, kicking should be expected and additional restraint such as the tail hold, the flank fold grip or the antikicking clamp may be needed.

References

Chesterton, R. N., Pfeiffer, D. U., Morris, R. S. and Tanner, C. M. (1989) Environmental and behavioural factors affecting the incidence of foot lameness in New Zealand dairy herds–a case-control study. *New Zealand Veterinary Journal* **37,** 135–142.

Fowler, M. E. (1978) Cattle and other domestic bovids. Chapter 9 in *Restraint and Handling of Wild and Domestic Animals*, pp 113–130. Iowa State University Press, Ames.

Grandin, T. (1980) Livestock behavior as related to handling facility design. *International Journal for the Study of Animal Problems* **1,** 33–52.

Holmes, R. J. (1984) *Sheep and Cattle Handling Skills.* Accident Compensation Corporation, Wellington (NZ). 36 pp.

Holmes, R. J. (in press) Safer practices in handling farm animals. *Proceedings of Farmsafe 88.* University of New England Press, Armidale (NSW, Australia).

Leahy, J. R. and Barrow, P. (1953) *Restraint of Animals*, 2nd edition. Cornell Campus Store, Ithaca (NY).

Miller, W. C. and Robertson, E. D. S. (1959) Casting animals as a method of restraint. In: *Practical Animal Husbandry*, 7th edition, pp 79–104. Oliver & Boyd, Edinburgh.

Seabrook, M. F. (1972) A study to determine the influence of the herdsman's personality on milk yield. *Journal of Agricultural Labour Science* **1,** 45–59.

Stöber, M. (1979) Handling cattle: calming by means of mechanical restraint. In: *Clinical Examination of Cattle*, 2nd edition (Editor G. Rosenberger, translator R. Mack), pp 1–28. Paul Parey, Berlin.

Further Reading

Battaglia, R. A. and Mayrose, V. B. (1981) Livestock restraint techniques. Chapter 1 in *Handbook of Livestock Management Techniques* (Editors R. A. Battaglia and V. B. Mayrose), pp 1–62. Burgess, Minneapolis.

Clutton-Brock, J. (1981) Cattle. Chapter 6 in *Domesticated Animals from Early Times* (Editor J. Clutton-Brock), pp 62–70. Heinemann and British Museum (Natural History), London.

Fraser, A. F. and Broom, D. F. (1990) *Farm Animal Behaviour and Welfare*, 3rd edition. Baillière Tindall, London.

Grandin, T. (1987) Animal handling. *Veterinary Clinics of North America: Food Animal Practice* **3**, 323–338.

Johnston, B. and Gahan, B. (*c.* 1986) *Handling Cattle from Farm to Abattoir.* Department of Agriculture New South Wales, Sydney. 21 pp.

Kilgour, R. and Dalton, C. (1984) Cattle. Chapter 2 in *Livestock Behaviour: a practical guide*, pp 7–53. Granada, St Albans (UK).

Leaver, J. D. (1988) Dairy Cattle. Chapter 2 in *Management and Welfare of Farm Animals: the UFAW Handbook*, 3rd edition, pp 14–45. Baillière Tindall, London.

Lemenager, R. P. (1981) Beef cattle management techniques. Chapter 2 in *Handbook of Livestock Management Techniques* (Editors R. A. Battaglia and V. B. Mayrose), pp 65–102. Burgess, Minneapolis.

Lemenager, R. P. and Moeller, N. J. (1981) Cattle management techniques. Chapter 3 in *Handbook of Livestock Management Techniques* (Editors R. A. Battaglia and V. B. Mayrose), pp 105–181. Burgess, Minneapolis.

Moeller, N. J. (1981) Dairy cattle management techniques. Chapter 3 in *Handbook of Livestock Management Techniques* (Editors R. A. Battaglia and V. B. Mayrose), pp 183–210. Burgess, Minneapolis.

Seabrook, M., editor (1987) *The Role of the Stockman in Livestock Productivity and Management.* Report EUR 10982 EN. Commission of the European Communities, Luxemburg.

Webster, A. J. F. (1988) Beef cattle and veal calves. Chapter 3 in *Management and Welfare of Farm Animals: the UFAW Handbook*, 3rd edition, pp 47–79. Baillière Tindall, London.

3

Sheep

ROBERT J. HOLMES

Introduction

The purpose of this chapter is to describe how sheep can be handled in a humane, safe and efficient way. Skill in handling is best acquired by practical experience with a good shepherd. Some of the basic handling skills discussed in the chapter on cattle are relevant to sheep handling and are not repeated here.

Sheep are deceptive in their ability to damage people. Because of their size we tend to lift and move them manually, often with a bent back. Such actions may cause low-grade back injury and the effect is cumulative. This damage may not be felt for many years, but may, nevertheless, be permanent. It accounts for the high numbers of sheep farmers (and other manual lifters) who have to change occupation because of 'bad backs' (often the result of prolapses of intervertebral discs).

Although sheep are not generally aggressive towards people, they may cause injuries. Rams, particularly during the mating season, may attack with sufficient force to cause serious leg or back injury. Under strong provocation, such as being mustered with dogs, sheep may stampede, break through fences, knock people down, and run over them.

Species Characteristics

Sheep are a very social genus of the 'follower' type. From the first hour after birth, when they start walking, lambs move towards and follow large moving objects, which are usually their dams. Adult animals also readily follow people who have fed them as lambs.

They normally maintain visual contact with at least one other sheep. If they lose contact they immediately try to restore it. If this is not possible, they call loudly and try vigorously to rejoin other sheep. An isolated sheep may run or jump at, and even knock down, a person between it and other sheep, so it is a good idea to watch them at all times. They should, whenever possible, be kept within touch or sight of one another so that they are not aroused by isolation.

Sheep that are kept in restricted spaces like fields learn to respond to dogs by grouping. Feral sheep, on the other hand, scatter when attempts are made to muster them with dogs. Once domestic sheep are in a mob they try to keep a minimum distance (**flight distance**) from the dog and so move away on the dog's approach. Flight distance is shown by single sheep and groups. When they cannot escape, they are likely to turn and face, or run past, the dog to escape from the confinement.

Sheep do not sense their environment in the same way as man. They have eyes at the sides of their heads and can see almost right around themselves (Fig. 3.1) without turning their heads. They have a 'blind spot' of about 30 degrees directly behind. Sheep have a narrower field of stereoscopic vision, which gives depth perception, than humans. They do not have depth perception to the sides of their heads where they can only see out of one eye on each side. This explains why unfamiliar objects suddenly appearing in their side vision are likely to startle them.

As a species which is preyed upon, sheep monitor their environment continuously for potential danger and so panoramic vision is important to them. While they will react to a moving dog at up to 1000 m, they will not react to one which is still, as they cannot see as much detail as humans can.

Rumination occurs only under relaxed conditions when arousal is low.

Arousal

The concept of arousal — the degree of activity of an animal — and principles of its control are discussed in the cattle chapter. Common

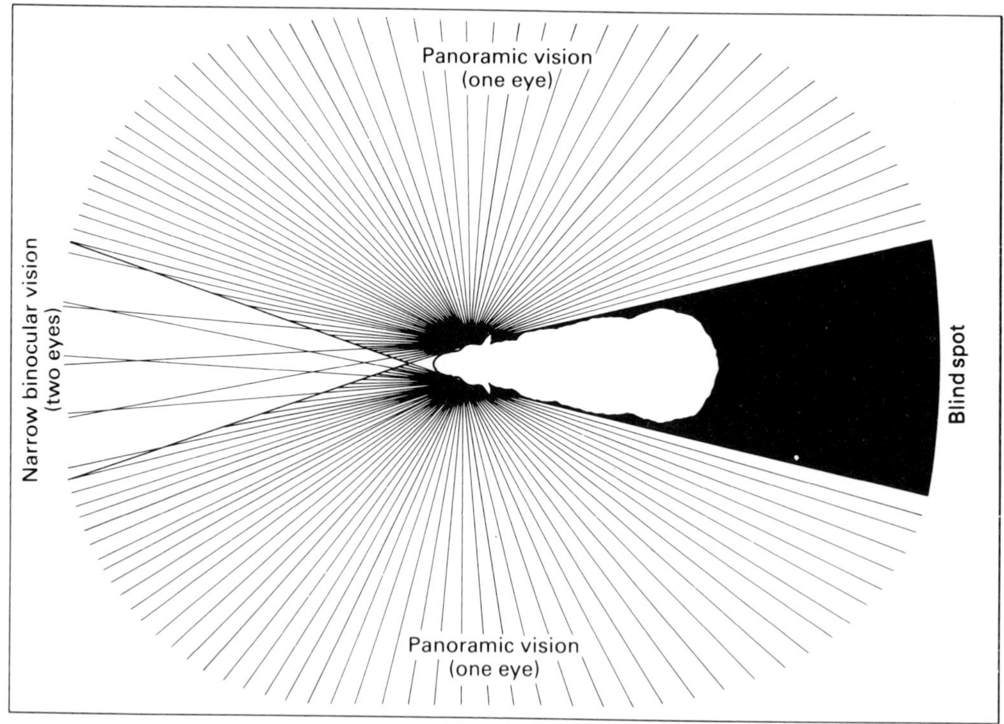

FIG. 3.1 The field of vision of a sheep.

observation shows that the following increase the arousal of sheep:

—visual isolation from other sheep
—proximity of dogs or people
—hunger or thirst
—sexual activity
—new environments
—unusual, loud, or high-frequency sounds;

and the following decrease arousal:

—proximity of other sheep
—low-frequency noises such as the 'rumble' call of a ewe to her lamb, or the courting ram to an oestrous ewe
—low illumination, as in covered yards.

Preparation

The benefits of planning to animal handling are described at greater length in the cattle chapter. Before starting any handling it is wise to have thought through all the tasks and possible responses of the sheep. Are all the facilities, tools, chemicals, and protective clothing ready? Are there sufficient people of the necessary skill? Is the route ready, with all unnecessary avenues blocked off? Once handling has started, such preparation helps the job to go smoothly and lessens delays due to unforeseen events.

General Control

Preparing

The route from field to yards should be prepared beforehand. All possible avenues of escape should be blocked off and the way cleared of all obstacles. Ideally, the group will move like a thick liquid, steadily and without turbulence, from start to finish.

Grouping

As mentioned before, domestic sheep group when a dog is in proximity and they may do this in response to a person. Once collected together they are much easier to drive.

Group Movement

It is much easier to move a group of sheep than to try to separate one or a few from a flock. It is sometimes more efficient to sort (draft or cut) a small group from a large flock using a gate. A well-trained dog or another person are essential and a flag is helpful to do such a draft. Flags can be used as described for cattle. It is often easier to take the whole group to the yards when just a few are needed.

Driving

Sheep can easily be driven by dogs and usually by people, because they try to maintain their flight distance. The position, posture, movement, and noise of dogs or people all affect the responses of sheep.

In warm weather it is best to start early and work in the cooler part of the day as the sheep are easier to handle — apart from it being more comfortable for the handlers.

Dogs and horses are probably the most effective aids for mustering or gathering, although some shepherds prefer motorcycles. To contribute their best, dogs need to be well-trained, worked regularly, and kept under control at all times.

There is a high risk of smothering when a large group is run through a narrow gap, such as through a gate in a rotational grazing system. The risk is reduced by:

—letting sheep move at their own pace
—moving the group uphill where possible
—placing a person immediately before the gap to cut off the flow if the smothering risk gets too high
—being particularly careful when a downhill movement leads to a stream crossing, culvert, gate, or other narrow area.

When driving a group up to a gateway, pressures such as dogs and noise should be eased to let the sheep look ahead to see where they are going.

The ease of getting to the yards depends a lot on the layout of fences, gates, and lanes. Ideally there should be no abrupt changes of direction, so corners should be rounded. It can be a good investment to build a lane or lanes feeding into the yards from distant fields.

To get a group into a small space or cul-de-sac it is best to keep the pressure off until the last possible moment and then put the pressure on until the movement is complete. The pressure should work first time, as once the sheep have broken back it will be more difficult next time. Getting them in is much easier if they can see at least 3 m of space through the end of the cul-de-sac. They can be driven forward by flapping or banging objects such as raincoats, truck inner tubes or canvas fire hoses on the ground. An effective, cheap and humane aid to forward movement is a rattle shaken behind the group. Small pebbles in a container, a number of can lids on a thick wire, or metal washers on a rod are all effective. However, with frequent use the animals become habituated to such aids.

When there is too much pressure, sheep mill around and climb on the backs of others. This interrupts the smooth flow through gaps. Once the flow has started, pull back on the pressure and let them follow 'just like sheep'.

When driving sheep along public roads, have one person at the front to warn oncoming traffic and to control rate of movement, and the more experienced person behind to move them forward.

Leading

Because they are 'follower' animals, sheep will naturally follow other sheep. They will also follow people who have fed them regularly, such as owners of pet lambs or people who regularly feed supplements. In some circumstances it is worth the time training decoy (leader) sheep.

Moving in Yards

Yard design has a major effect on the efficiency with which animals can be handled. Yards should be built to get the sheep to flow readily and to allow people to move easily around the yards and sheep. Good design takes account of what the sheep sees, hears, and smells, and how it reacts to those stimuli. Vision of sheep is the most important factor affecting the working of yards.

Yards can be designed so that sheep flow through them with very little stimulation, just the presence of a person behind the sheep. A well-controlled dog can be useful to a solo handler for such things as drafting when controlled pressure is required.

Dogs can be counter-productive in yards as they may be too strong a stimulus for the circumstances. They can cause the sheep to run away, run into structures, bruise themselves, jam in narrow gaps, fall over, climb on the backs of others, dislocate joints, or break limbs. The tendency of dogs to bite needs careful control.

If you need more than a quiet dog and a rattle to move sheep through the yards, it may be worthwhile thinking about improving the yard design.

Sheep will:

—move faster on the level than uphill, and uphill faster than downhill
—move faster with covered sides to the races
—usually stop and investigate changes in the environment, when in front of a moving group
—look down at the ground immediately after stopping if they have sufficient room to move their heads
—draw back from the edge of a drop ('visual cliff')
—baulk when they see there is a drop beneath the ground surface, such as under gratings
—stop, turn, and walk away at about 3 m from a dead end
—not readily walk or jump into water which is deep or of unknown depth
—attempt to force their way into a smaller gap than their body width, when fleeing.

Recent improvement for moving and restraining sheep include elevated delivery races, forcing pens, non-return gates, restraining conveyors, and handling machines. When putting sheep through yards for the first time, it is a good idea to mix them with sheep that have gone through before in order to reduce arousal.

Sorting

Efficient sorting (drafting, shedding, or cutting) depends on a steady stream of sheep moving up a single-file race to one or two swinging gates. Suitable pressure is needed behind the sheep. A solo handler will need a well controlled dog. Where

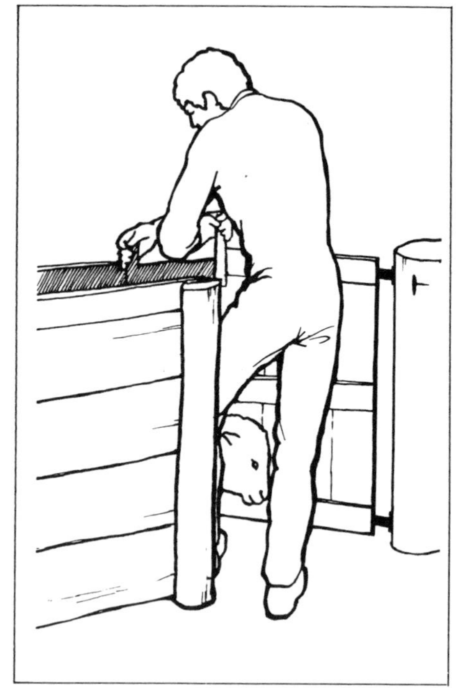

Fig. 3.2 Blocking sheep during sorting using the leg closest to the sheep.

there is another person, many shepherds tie their dogs up outside the yards well out of the way. The assistant can apply pressure by shouting, flapping or rattling objects, as described above.

Anything that goes into the race — such as your knee or groin — is likely to be hit head-on by a fast-moving sheep. Sometimes lambs leap headlong into the sorter's body. It is better to use the side of the leg (Fig. 3.2) to block a sheep. This is done by standing facing across the race and using the leg nearest the oncoming sheep to block. The knee is less likely to be injured by a blow to the side, and the side of the body is presented to the sheep instead of the front.

Catching

Stout boots are preferred to rubber boots or light shoes when working close to sheep. By reducing concern for trampled toes, they make it easier to concentrate on the task in hand.

It is easiest and safest to catch sheep by approaching them from behind (Fig. 3.3) in the 'blind spot'. They are most easily caught in a pen where there is just enough room for the handler to

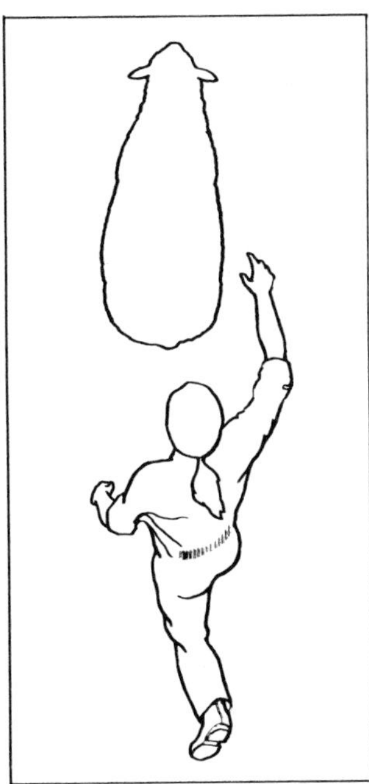

FIG. 3.3 Catching a sheep from behind in the 'blind spot'.

move around freely but not enough for the sheep to do so. If the pen is too big for that, they can be bunched up into a corner.

There are several ways of catching, but the most effective and safest for the sheep is to restrain the head. This can either be done by cupping the hand under the chin, or by catching the neck in a crook (Fig. 3.4). Holding by the wool is a bad habit to develop as it causes skin bruising ('wool pull'), which damages the carcase and results in downgrading.

FIG. 3.4 Catching a sheep by a crook around the neck.

FIG. 3.5 Restraining a sheep by holding against the pen side with arms and knees.

An animal with horns can be difficult and dangerous to catch by cupping the hand under the chin. It can be caught by a neck crook. Another way is to grab and lift a hind leg but you will need a strong arm and back for larger sheep. Once it has stopped struggling, transfer your grip to one and then both horns. The use of leg crooks is not recommended as they occasionally cause udder and leg damage.

Holding

It is easiest to keep a standing sheep still by holding it against the pen side with arms and knees (Fig. 3.5).

In an open area where there is no fence and the sheep is small enough, it may be straddled, as shown in Fig. 3.6, with one hand under the chin, the back straight and standing as far back as is comfortable.

A standing sheep can be held in a neck yoke as shown in Fig. 3.7.

Lifting

Many sheep farmers suffer 'bad backs' each year and lifting sheep is one of the most common causes.

FIG. 3.6 Holding a sheep in an open space.

Sheep are heavy, difficult to hold against your chest, likely to struggle and have no convenient places to grip. All these factors increase the risk of back injury (see above). If possible, gates or ramps should be used instead of lifting sheep.

If lifting is unavoidable:

—do warm-up exercises
—keep a straight back
—bend the knees and use the legs to lift
—lift smoothly and continuously
—know your limits.

The recommended method for lifting a sheep up into a vehicle or over a fence is illustrated in Fig. 3.8.

1. Hold the sheep against the structure and straddle the rump. One hand should be on the structure for support, the other under the sheep's neck.
2. Using the support to take the strain off the back, pull the sheep's forequarters up so that it is standing on its hind legs.
3. Move the hand from the support under the nearest leg to grasp the opposite front leg.
4. Move the other hand from under the neck to grasp the flank fold of skin firmly.
5. Bend the knees to crouch behind the sheep and take the sheep's weight on the thighs.
6. Hold the sheep firmly.
7. Stand up to lift the sheep off the ground and, in one continuous movement, swing the sheep upwards towards the structure, giving it an extra lift with your knee.
8. If lifting over a rail, put the sheep on top of the rail and let it see the ground.
9. With a rolling movement let the sheep fall on to its feet.

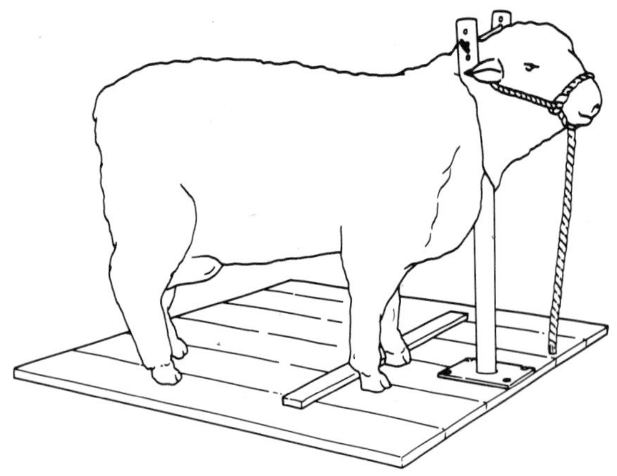

FIG. 3.7 Sheep held in neck yoke.

The whole procedure needs to be done smoothly. For a taller sheep, start by standing between the rail and the sheep. If in doubt about your ability to lift — don't.

Walking

From the position described above for holding a small sheep in an open area (Fig. 3.6), the sheep's movement can be controlled by moving the body backwards, or forwards from the shoulder. By standing further back and, if necessary, squeezing with the knees, the sheep can be made to move forward. Squeezing the tail-head also makes it move forward.

Bigger sheep can be moved using the position shown in Fig. 3.9. Forward movement is limited by holding the sheep back with the hand cupping the neck and, if necessary, turning the sheep's head and neck. It can be made to go forward by squeezing the tail-head.

A sheep can be walked backwards on its hind legs, as is often done by shearers, rather than dragging the sheep from the catching pen to the shearing board (Fig. 3.10). The sheep is lifted up to stand on its hind legs in the same way as for the start of lifting (Fig. 3.8). Once on its hind legs it can be held against the chest by holding the forelegs and walking backwards.

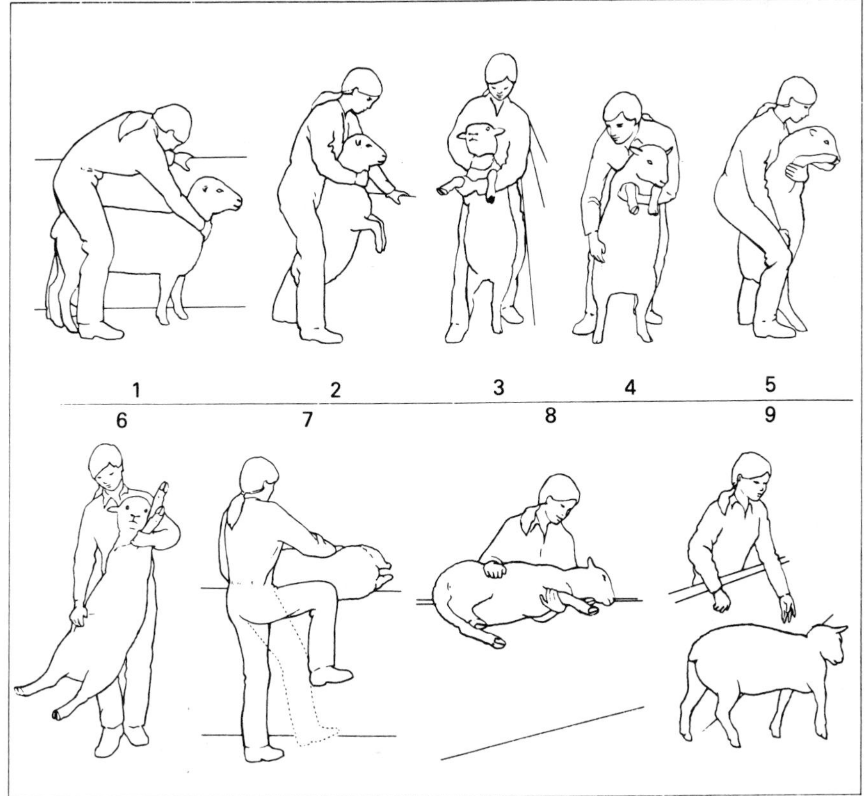

FIG. 3.8 Lifting a sheep over a fence.

FIG. 3.9 Controlled walking of larger sheep.

Sitting on the Rump

To hold a sheep for more than a few seconds it is advisable to sit it on its rump. Once in this position sheep usually become quiet, although they may struggle for a few seconds at first.

It is not necessary to lift a sheep off the ground to sit it on its rump. The recommended method where there is sufficient space is shown in Fig. 3.11. The sheep is held against braced knees with one hand under the chin and the other on the rump. The head is turned to face the rear and the hindquarters forced down against the leg. When the sheep is no longer standing on its hind legs, the front end is lifted up with the back straight and it is positioned on its rump. It will sit better if it is sat on one side of its hindquarters. The sheep can be leaned against you and held more easily with the toes in, the feet spread slightly, and the knees together.

If there is not enough room for this method, the

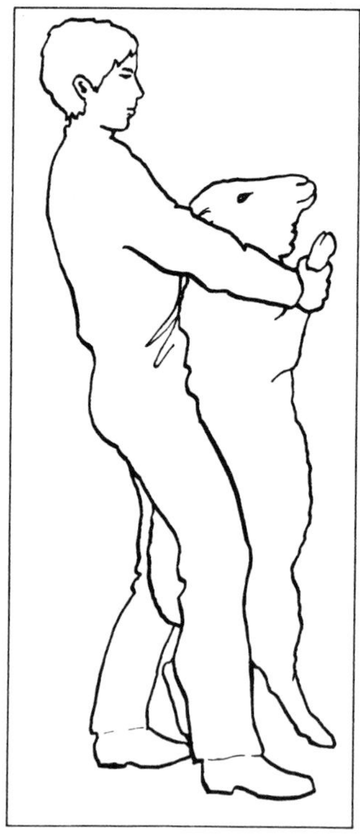

FIG. 3.10 Walking a sheep backwards.

fore-end can be lifted as shown in Fig. 3.8. With the sheep standing on its hind legs only, both forelegs are held and pulled against one's chest. By pressing one's knees forward into the sheep's back, its legs can be pushed from under it so that it sits on its rump.

Most, but not all sheep, are quiet in this position. Others may struggle for a little while. Twisting the sheep's head down and to the side usually makes it quieter.

Immobilizing

Numerous techniques, using varying degrees of physical restraint and positions, are effective for keeping sheep still for short periods. It is vital that any method used for more than a few minutes allows the gases from the rumen to escape. Otherwise the sheep may die from bloat. Sitting a sheep on its rump as described above usually keeps it still for a few minutes.

Tying up

A sheep can be kept lying on the ground by restraining either the fore or hind legs. A cheap and effective method for tying up the hind legs with a loop of rope or twine is shown in Fig. 3.12.

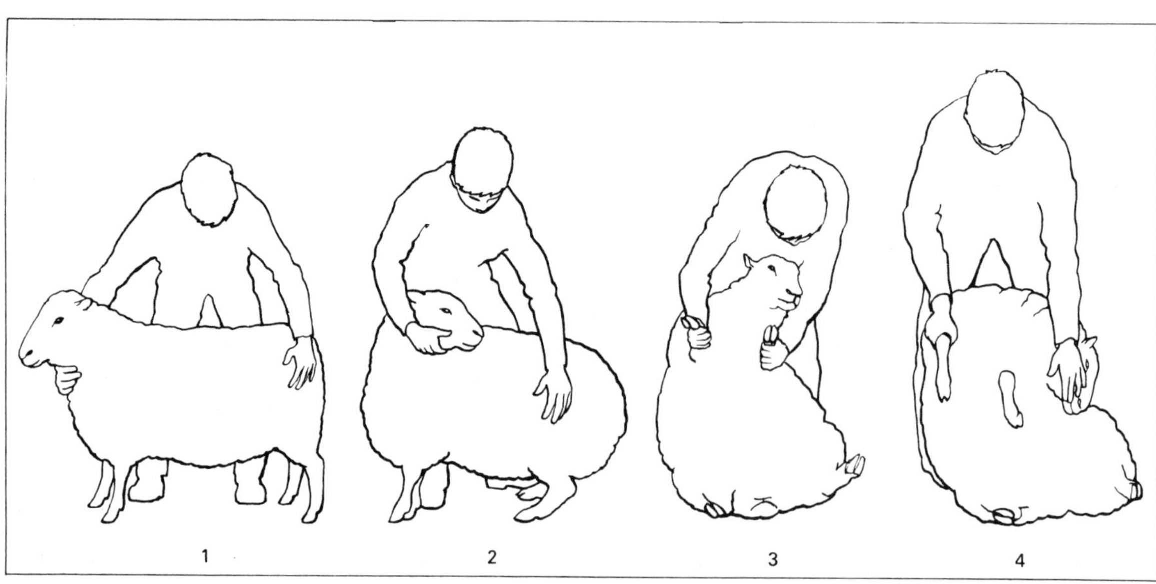

FIG. 3.11 Turning the head of a sheep to sit it on its rump.

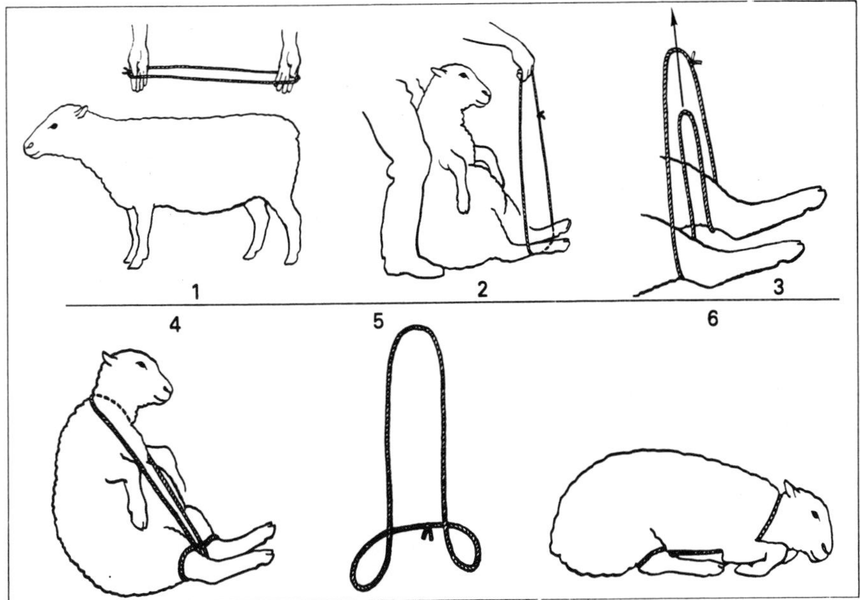

FIG. 3.12 Tying up a sheep with a loop.

FIG. 3.13 Sheep immobilized on its chest by a gambrel.

With the sheep sitting on its rump, a loop the length of the shoulder-to-rump distance is dropped over the hind legs and placed above the hocks. The part of the rope between the hind legs is pulled upwards to make two small loops and one large one. The small loops encircle each hind leg above the hock and the large loop goes over the neck. The sheep is laid on the ground on its sternum. The length of loop may need adjusting so that the sheep can lie comfortably but not stand. On a slope, the sheep's head should point uphill to allow escape of rumen gases.

Gambrel

A plastic gambrel is commercially available. The device is curved and fits over the neck of a sheep lying on its chest (Fig. 3.13). The forefeet are individually lifted up and placed in crooks on top of the gambrel. The forelegs exert downward pressure on the gambrel which holds the neck down.

Handling machines

There are also handling machines available for paring feet and other tasks. By various means they usually tip the animal onto its back and clamp it in position. There is also a restraining conveyor for presenting sheep in a controlled way to the place of manipulation. Sheep walk into a 'V'-shaped structure composed of two conveyor belts. When the sheep have walked sufficiently far in, the floor surface falls away and the sheep is carried forward in the 'V'. These have proved very successful for presenting sheep for stunning before slaughter and for foot paring. When a sheep reaches an operator, its head is lifted up and the animal is tipped over on to its back by the motion of the conveyors. The sheep is then cast within the conveyor.

Tonic immobility

Restraint and inversion is a common way of causing an hypnotic state, which is called tonic immobility, in many species. Sitting a sheep on its

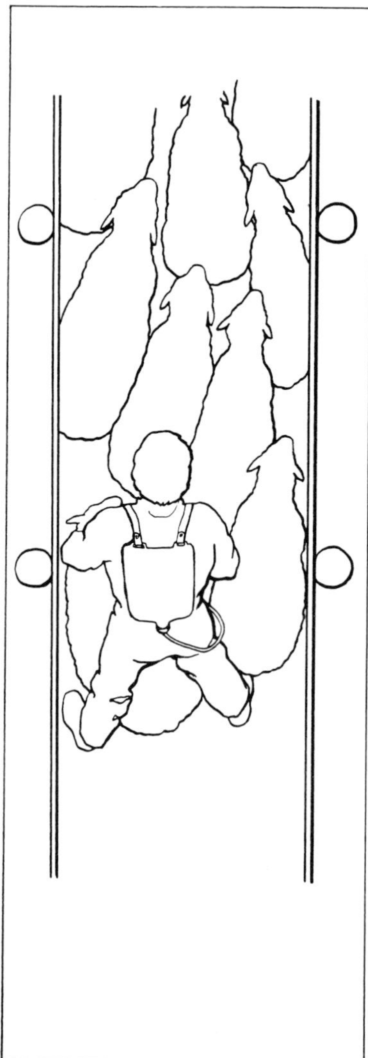

Fig. 3.14 Drenching race.

rump or holding it on its side on the ground causes this state in most sheep. Various methods of causing and prolonging the state have been tested. To date, the most effective way has been to restrain the sheep on its side, wrap a cuff around the ears and put a box over its head. Putting a hat over a sheep's eyes usually keeps it lying on its side for a few minutes.

Common Manipulations

Weighing

Newly-born lambs can be weighed in a sling beneath a hand-held scale. This is done in some flocks as soon after birth as possible to reduce the chances of error in pedigree recording. At whatever age, it is often done at the same time as antiseptic treatment of the navel, ear-tagging, castration and tailing.

Once lambs are too heavy to lift conveniently for weighing they are put through a race containing an individual weighing cage. It is a good idea to have sheep visible in front of and behind the cage until the animals are used to the procedure, as they are likely to be highly agitated by the separation.

Drenching

A group of sheep can be most easily drenched (dosed) when tightly packed and facing the same direction in a race about 1 m wide (Fig. 3.14). Start at the back and move forwards through the group.

A drench gun with a short nozzle and thick rounded end is used (Fig. 3.15). The nozzle should be smooth so that the mouth is not injured.

The head is held lightly under the chin with one hand and the nozzle inserted from the side of the mouth behind the incisors but in front of the molars (Fig. 3.15). The fingers need not be put in the mouth, thus avoiding the risk of damage. The head should be kept as horizontal as possible. Lifting the head up makes it harder for the sheep to swallow and increases the chance of damaging the roof of the mouth with the nozzle.

The nozzle is inserted into the middle of the mouth, taking care it does not get between the molar teeth (Fig. 3.15). Be careful also to avoid knocking the teeth as this hurts the sheep and makes it more difficult to handle next time.

When the nozzle is over the back of the tongue, gently squeeze the trigger. It is important to squeeze the trigger gently so that the drench is received slowly enough to be swallowed. If it is squirted violently, the drench is likely to do down the trachea and cause pneumonia and death.

Take the nozzle out of the mouth before letting go of the trigger. Some drench guns suck back liquid when the trigger is released.

Examining Incisors

Checking the mouth to see the number and state of the incisors can be conveniently done in the

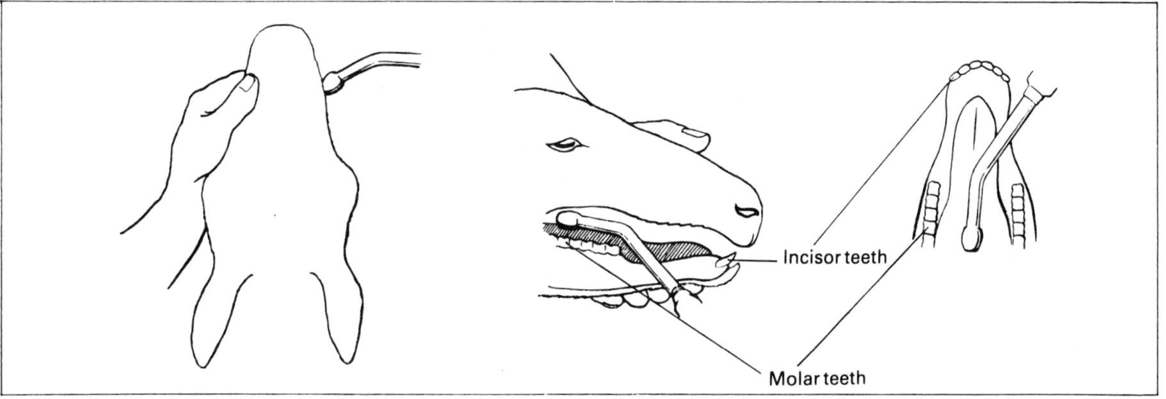

Fig. 3.15 Position of the hand and gun for drenching.

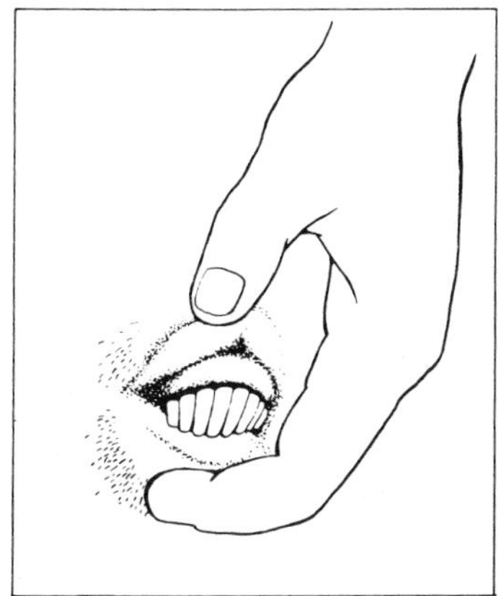

Fig. 3.16 Exposing the incisors.

same race as for drenching. A quick, simple, and safe way to see the incisors is to pull back the upper and lower lips simultaneously with your thumb and forefinger (Fig. 3.16). Avoid blocking the nostrils as stopping its breathing makes the sheep toss its head about and become difficult to handle around the head again.

While the incisors are exposed you can see how they meet the dental pad and whether the jaw is undershot, normal, or overshot.

Checking for loose incisors requires the mouth to be open. This is done by cupping the jaw with the fingers and pushing the thumb in the space between the incisors and molars. Looseness can be tested with the other hand.

References

Bremner, K. J., Braggins, J. B. and Kilgour, R. (1980) Training sheep as 'leaders' in abattoirs and farm sheep yards. *Proceedings of New Zealand Society of Animal Production* **40,** 111–116.

Holmes, R. J. (1980) Dogs in yards: friend or foe? *Quarter* **1**(3), 16–17. Ministry of Agriculture and Fisheries, Wellington (NZ).

Holmes, R. J. (1984) *Sheep and Cattle Handling Skills.* Accident Compensation Corporation, Wellington (NZ). 36 pp.

Holmes, R. J. (1987) Sheep behaviour and welfare in New Zealand slaughterhouses. *Annual Proceedings of Sheep Veterinary Society of British Veterinary Association* **12,** 106–114.

Further Reading

Battaglia, R. A. and Mayrose, V. B. (1981) Livestock restraint techniques. Chapter 1 in *Handbook of Livestock Management Techniques* (Editors R. A. Battaglia and V. B. Mayrose), pp 1–62. Burgess, Minneapolis.

Fowler, M. E. (1978) Sheep and goats. Chapter 8 in *Restraint and Handling of Wild and Domestic Animals*, pp 131–138. Iowa State University, Ames.

Fraser, A. F. and Broom, D. F. (1990) *Farm Animal Behaviour and Welfare*, 3rd edition. Baillière Tindall, London.

Kilgour, R. and Dalton, C. (1984) Sheep. Chapter 3 in *Livestock Behaviour: a practical guide,* pp 54–84. Granada, St Albans (UK).

Outhouse, J. B. (1981) Sheep management techniques. Chapter 7 in *Handbook of Livestock Management Techniques* (Editors R. A. Battaglia and V. B. Mayrose), pp 354–408. Burgess, Minneapolis.

Williams, H. Ll. (1988) Sheep. Chapter 4 in *Management and Welfare of Farm Animals: the UFAW Handbook*, 3rd edition. Baillière Tindall, London.

4

Goats

ALASTAIR R. MEWS and ALAN MOWLEM

Introduction

Goats were one of the first species of mammal to be domesticated by man. Perhaps only the dog has been domesticated longer. It is relevant to this chapter to understand that goats probably domesticated themselves. They are very inquisitive animals and almost certainly wandered into the camps of early man out of sheer curiosity.

There are two main interests in goats in the United Kingdom. There is a long established tradition of goats being kept as a hobby and a keen interest in pedigree breeding. The British Goat Society, the pedigree registering authority in the United Kingdom, has been established for 110 years. In more recent years there has been a growing interest in goat farming and goats are now kept in large numbers on some farms, in some cases as many as 500.

There are two types of goat kept in the United Kingdom: dairy goats for milk production and goats kept for fibre production. Several of the dairy breeds, such as the Saanen, Toggenburg and British Alpine, originate from Switzerland. The one other well known breed, the Anglo Nubian, with its characteristic large nose and droopy, long ears, is the result of crossing Egyptian and Indian breeds with indigenous English Goats at about the turn of the century.

The goats kept for their fibre are the Angoras, originating from Turkey, which produce mohair, and the cashmere-producing goats. Cashmere is the soft underdown produced for insulation by a number of breed types kept in cold environments.

Handling goats is very much influenced by the fact that they seem to enjoy companionship and being handled, once they have become used to it. Regularly handled dairy goats are often embarrassingly friendly and can be quite overwhelming when in groups. Fibre-producing breeds seem to be more reserved, but even they soon get used to being handled, and in small numbers can be almost as amenable as the dairy breeds. The exceptions are the males. Although rarely unfriendly, their strong, almost overpowering, smell makes them most unattractive animals to handle and as a result they are often handled badly.

Catching and Restraint

The inquisitiveness of goats can often be used to good advantage when catching individual animals. The group should be approached quietly and in a friendly manner. The handler should enter the pen and, if necessary, pretend to carry out some task to attract the attention of the goats. They will then almost certainly crowd round and it is then a simple matter to hold the individual required. Although very inquisitive, they are also cunning and soon learn when someone is trying to catch one of them. It is therefore preferable and time saving if they are caught without recourse to a chase as this can often be very unproductive.

Once goats are caught, they can be restrained by a firm hold around the neck at the base of the skull. If a difficult procedure is to be carried out, a collar should be used. A dog collar of the size used for a Labrador is suitable for an adult goat. Fibre-producing goats such as Angoras should not be restrained by holding their fleece as this will cause excessive bruising under the skin.

Goats that are not used to regular handling, such as some fibre-producing herds run out on hill land, will need to be handled in a race for routine treatment such as vaccinations and drenching. Some goats are expert jumpers and escapees, and therefore the handling race must be soundly constructed with a minimum height of 1.5–2 m.

Goats are natural followers and should, wherever possible, be led rather than driven. A lot of time can be saved if the goats are taught to

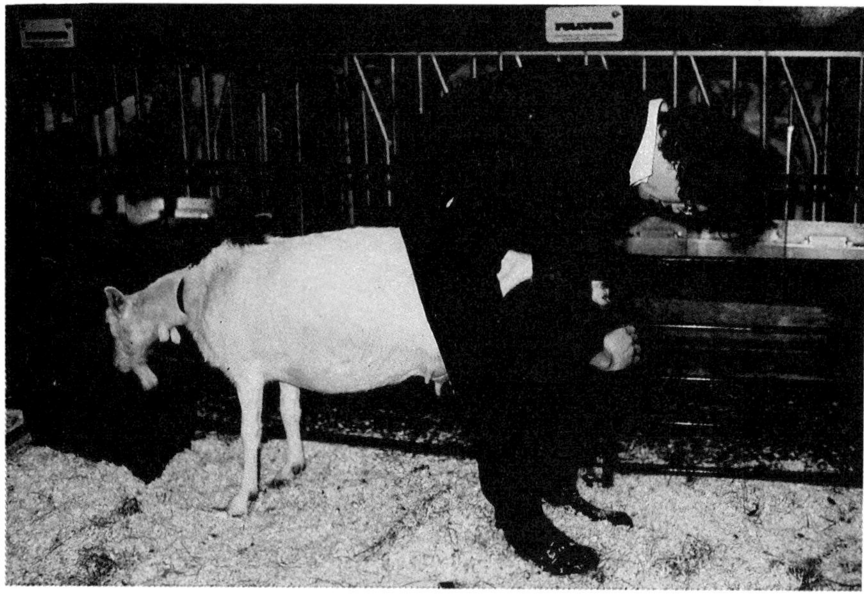

FIG. 4.1 Foot-trimming.

recognize a food bucket or bag. They will usually follow a handler into a pen if they think they are going to be fed.

Manipulations

Foot-trimming

With one exception this is always carried out with the goat in a standing position (Fig. 4.1), if necessary tethered to a gate or pen-side. Its feet are held in a similar position to that for shoeing a horse. If the goat is at all fractious, the situation will usually be improved if it is held by a second person, rather than being tethered. The exception is the Angora goat. These are more 'sheep-shaped' and are covered with a dense fleece which allows them to be cast like sheep and restrained in a sitting position with their backs against the handler's legs and knees. They are not quite as broad as most sheep and therefore are more comfortable if they are inclined to one side, so that their weight rests on their upper thigh rather than the centre of their spine.

Drenching

The position and restraint are exactly the same as those used for sheep, with the head inclined to one side and the drenching gun or bottle being inserted in the side of the mouth to pass through the diastema. Maldrenching is unusual but, nevertheless, care should be taken to ensure that the drench does not go down the trachea.

Injections

If carrying out injections single-handed, the goat should be tethered as described for foot-trimming although, ideally, another person should restrain it while injections are being carried out. The usual site for subcutaneous injections is the lower neck or behind the shoulder. Intramuscular injections are normally given into the gluteal muscle. Goats tend to be less well muscled than sheep and care should be taken to make sure the needle does not go through the muscle into the periosteum. Intravenous injections are normally given into the jugular vein. If the goat has tassels or wattles, the jugular vein runs vertically down from these. In all but heavily fleeced Angora goats, it is easily located if it is occluded with a thumb and finger at the base of the neck, with the head turned slightly away. For euthanasia of kids using barbiturates, intracardiac injection may be considered preferable as the comparatively small jugular of a kid may be difficult to find.

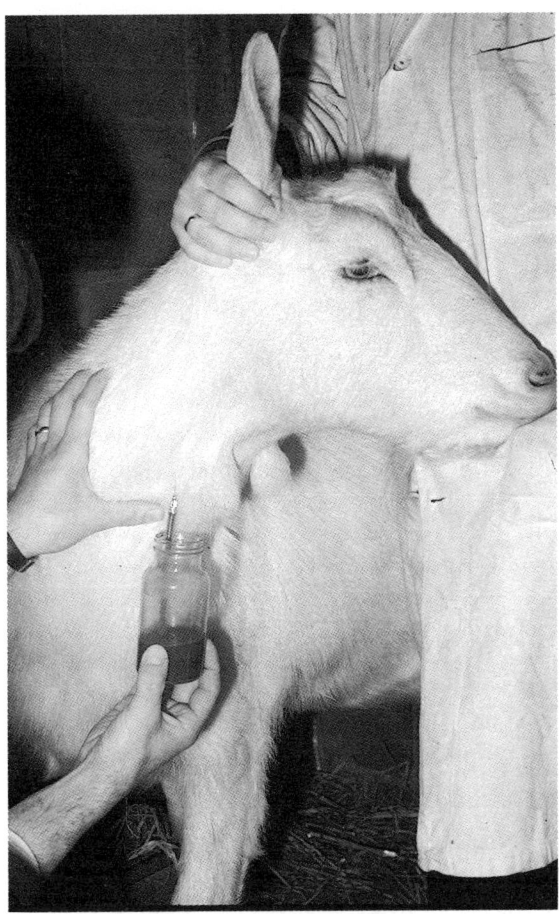

FIG. 4.2 Collecting a blood sample from the jugular vein.

Some pedigree breeders who show their goats will be very particular about vaccination sites because of the problem of reaction to some vaccines and the possibility of consequent injection-site abcesses.

Infusion of Therapeutic Agents into the Udder

A wide range of antibiotic preparations is available in tubes and applicators with narrow nozzles for this purpose. As most of these preparations are indicated for cows there is some confusion about dose levels for goats. However, the goat udder half is of similar capacity to a cow udder quarter and the teat and gland cistern are proportionally larger and therefore the same dose level is normally used. Milk should be stripped out from the udder before infusing antibiotics which should be massaged well into the udder.

Restraint for General Anaesthesia

If a goat is to be given a general anaesthetic, it should be handled gently and firmly with, wherever possible, someone known to the goat to calm and reassure it. Goats have less muscle and fat cover than some other animals and care should be taken to ensure that bruising is avoided when they are laid down and restrained. For surgery it is preferable to use a gaseous anaesthetic administered via an endotracheal tube. A size 10–12 tube should be used for an adult goat. The tube and the head are held lower than the abdomen to prevent saliva or rumen contents from entering the lungs.

Disbudding and Dehorning

Kids should be disbudded preferably within four days of birth. General, rather than local, anaesthesia is preferred as the latter is difficult to ensure in the kid. In addition, overdosing can easily occur. Once again kids should be laid down on a cloth or similar soft material and the head firmly restrained to facilitate the application of the disbudding iron.

The procedure for restraining a goat for dehorning is the same as described for major surgery using a general anaesthetic. It is a traumatic operation for an adult goat and should only be carried out if the horns themselves are creating welfare problems. Full details of the methods for these procedures are described by Buttle *et al.* (1986).

Shearing

Two types of goats are kept for their fibre. Angoras produce mohair whilst a variety of long-haired types produce soft under-hair or down known as cashmere. Angora goats are restrained for shearing in much the same way as for sheep and as described for foot-trimming. Cashmere may be harvested by combing or shearing with the goat normally restrained in a standing position. Both types may suffer cold shock after shearing and therefore may need housing for at least a few days, depending on the weather and the time of the year.

FIG. 4.3 Shearing a male Angora.

FIG. 4.4 Collecting semen from a Bagot male.

Artificial Insemination

Techniques for semen collection and insemination are now well developed and present no particular problems (Mowlem, 1983). Semen is collected in an artificial vagina from a male which is encouraged, or trained, to mount a teaser female. With some males, particularly experienced ones, the teaser does not necessarily have to be in oestrus. It is best if the female is restrained in a crate (Fig. 4.4). This will prevent her from moving and making the job of the semen collector very much easier. A shortened bull artificial vagina is used. Semen can also be collected with the use of an electro-ejaculator, but this is a painful procedure and the quality of the semen collected is usually poor. It is therefore of dubious value with goats.

The preferred method for restraining the female to be inseminated is to hold her up with her head facing back between the handler's legs. Her rear end is raised up by lifting the bent back legs and she is presented to the inseminator (Fig. 4.5). This head-down, bottom-up position makes the inseminator's job of locating the cervix, using an illuminated speculum, very much easier. The insemination 'gun' is inserted into, but rarely through, the cervix.

Parturition

Anyone with experience of lambing sheep should have no problems adjusting to kidding goats. Dairy goats are prolific and therefore the chances of two or three kids being produced, is high. This increases the chances of malpresentation and sometimes the chances of kids becoming mixed up. Because of this, handlers assisting at birth must make sure that if they are applying traction, it is only to one kid. A recumbent newborn kid may sometimes be resuscitated using the same methods as for lambs. A dramatic, but often successful, method is to hold the kid by the hind

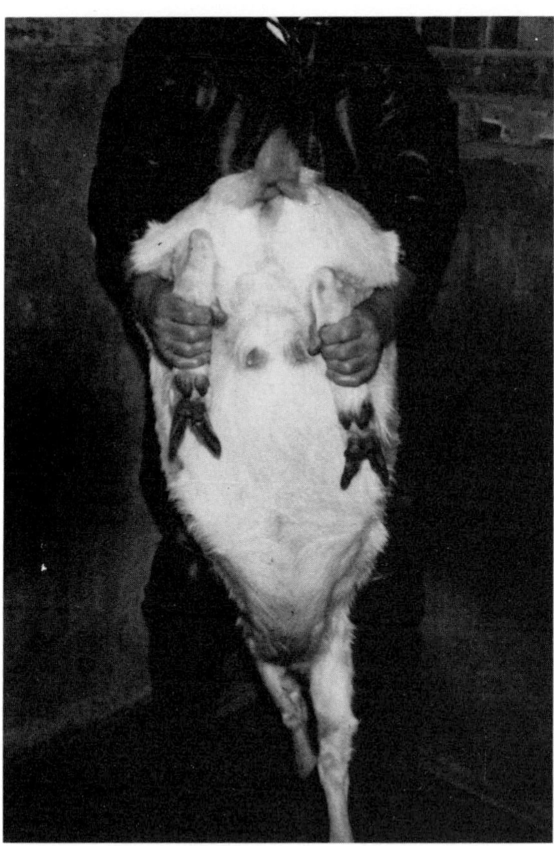

FIG. 4.5 Presenting a goat for artificial insemination.

legs and to swing it backwards and forwards or over and over, with the kid describing a circle. This forces fluid out of the lungs and is often enough to stimulate breathing. If new-born kids are too weak to suck their mother's teat, it will be necessary to feed colostrum via a stomach tube. These are sold for this purpose, complete with a receptacle for the colostrum, and are an important part of any lambing/kidding kit. Care should be taken to ensure that the tube is in the oesophagus and not in the trachea. Insertion of the tube is facilitated by holding the kid's head up with the neck extended.

Transport

If goats are being transported in a vehicle, they should not be allowed too much space, otherwise they will be knocked against the sides as the vehicle moves. If only a few are to be transported, spare space can be filled with bales of straw. These must be firmly tied in place and should not provide the goats with any climable surfaces or ledges. The ideal floor is a rubber mat covered with a thin layer of straw. Horned and dehorned goats should not be carried together in the same compartment. Goats should never be transported in a vehicle without a division between them and the driver.

References

Buttle, H., Mowlem, A. and Mews, A. (1986) Disbudding and dehorning goats. *In Practice* **8,** 63–65.

Mowlem, A. (1983) *Development of Goat Artificial Insemination in the UK. British Goat Society Yearbook.* British Goat Society, Bovey Tracey.

Further Reading

Dunn, P. (1987) *The Goatkeeper's Veterinary Book,* 2nd edn, Farming Press, Ipswich.

Guss, S. (1977) *Management and Disease of Dairy Goats,* Dairy Goat Publishing Corporation, Arizona, USA.

MacKenzie, D. (1980) *Goat Husbandry*, 4th edn. Faber and Faber, London.

Mowlem, A. (1988) *Goat Farming*. Farming Press, Ipswich.

Wilkinson, J. M. and Stark, B. A. (1987) *Commercial Goat Production*. BSP Professional Books, Oxford.

5

Deer

T. JOHN FLETCHER

Introduction

Since at least classical Roman times, deer have been enclosed within parks. They were kept, as Columella said, '. . . sometimes to serve for the magnificence and splendour and pleasures of their owners; and, at other times, to make gain and increase their revenue'. What distinguishes the modern deer farmer from the keeper of park deer is the farmer's ability to muster and handle his deer more or less at will. This enables him to manipulate the breeding of the animals in the purposeful, selective way required for complete domestication and the planned development of 'breeds' with desirable properties.

The design of the mediaeval deer parks of Europe, and especially Britain, came very close to that needed for deer handling by the present-day farmer. Some 2000 parks existed within England and Wales during this period and there were specialist deer keepers who probably travelled from park to park, designing and building deer pens or 'hays'. Thus, at Falkland Park in Fife, Scotland, Andrew Mathieson was paid to 'wynd' or weave a 'hayyard' or fenced pen and Master Levisay, an Englishman, was responsible for catching the deer without harming them. The deer captured in these pens were often moved by horse-drawn litter to stock other deer parks.

The weaving of deer pens to create a wattle fence represents the first important principle in handling deer: that the animals will not run at visually-opaque barriers. Thus, while high-tensile spring steel net fences are ideal for containing deer within paddocks, it is necessary to use timber rails, planks or, temporarily, even hessian or burlap fabric attached to the fence, when the animals are being moved into more confined areas for handling.

The tendency for deer to hurl themselves at less solid looking obstructions has been used by man for millenia in the capture of deer by 'netting' (Fig. 5.1). Indeed, some deer parks still use nets several hundred metres long to trap fallow deer (*Dama dama*). Nets may also be used by research workers to capture and mark wild populations of roe deer (*Capreolus capreolus*) and even in France, red deer (*Cervus elaphus*). Nets have, however, no place for the deer farmer; their use is labour-intensive, inconvenient, stressful and often injurious to the deer.

Handling Wild Deer

Before examining the techniques of handling farmed deer in detail, we should first look further at modern techniques used to capture wild deer.

Since deer farming on a large commercial scale began only as recently as the early 1970s, and since the breeding life of a hind is some fifteen years, many deer on farms started their life in the wild. This is even true of the million or so red deer hinds on New Zealand farms, since, until 1986, the high price of red deer female breeding stock made the capture of wild deer a more attractive commercial proposition. Although limited numbers were captured by feeding deer into farmed enclosures, it was the use of the helicopter that made the capture of around 15,000 deer a year possible. Originally they were tranquillized by dart guns, but this soon gave way to the use of a gun designed to shoot a net over the deer. The netted deer was then restrained with broad straps around its feet, often injected with a sedative, put into a canvas bag and rapidly flown to a farm for rehabilitation. The speed and efficiency of this system kept mortality to surprisingly low levels (1–5 per cent) and has proved highly effective, not only in stocking deer farms, but also in reducing the population of wild red deer, which had risen to such high levels as to seriously endanger the native bush.

FIG. 5.1 Netting a fallow buck.

In Spain, wild deer for farming are captured by driving, or feeding, into net corrals. The net is kept loose and, once entangled, the deer are lifted into crates. In Scotland, the over-population is such that wild deer are close to starvation during the winter. Therefore, in hard weather they can be readily fed into deer-fenced enclosures, from where they can be driven down a chute into pens for sorting before loading onto a lorry (Fig. 5.2).

Paradoxically, recently captured wild deer, once within suitably designed handling pens, are more easily managed than the larger farm bred animals in which familiarity has bred contempt for man.

Procedures which Require Handling of Farmed Deer

The ability to control the breeding of an animal in such a way as to produce domesticated strains is what distinguishes the farmer from the animal keeper. The modern farmer must be able to split his female stock into rutting groups and introduce the stags, all simple procedures which require facilities to draft the deer and pens to hold groups of up to 40 animals. In addition, he must have farm handling yards to enable him to determine which strains of deer he wishes to select, weighing facilities and pens in which eartagging can take place. Records will also be required of which hind has lactated. There must also be an area in which hinds can have their udders examined and records made of lactational status.

Other farm procedures which require handling include medication, which normally entails either oral administration (drenching), or injection of anthelminthics and vaccines and, less frequently, mineral supplements and antibiotics.

The farmer will almost certainly need to carry out tuberculin skin testing. This demands good lighting and sufficient restraint to allow very care-

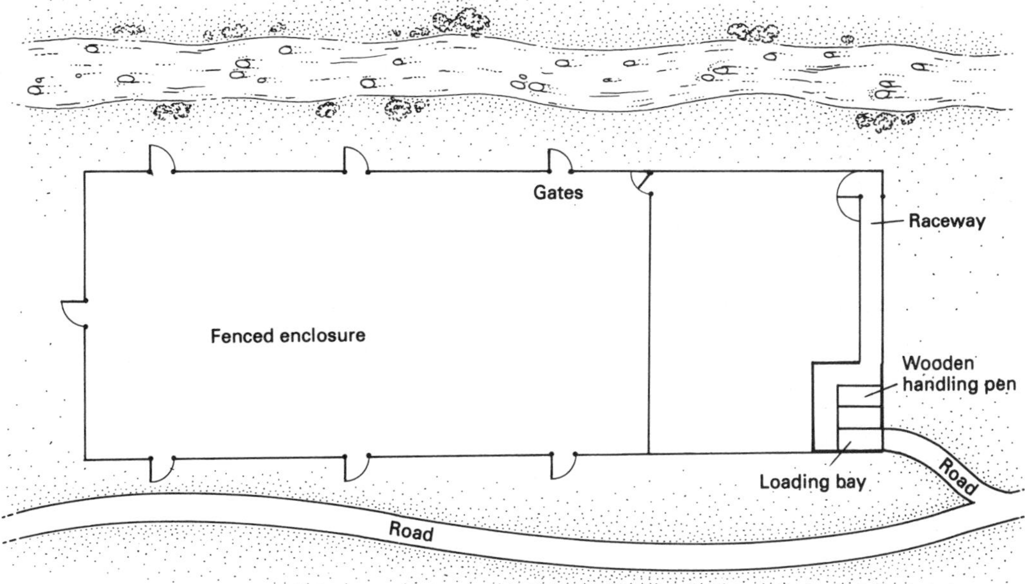

FIG. 5.2 Typical layout for a deer catching system in the bottom of a Scottish highland glen.

ful clipping of the injection site and measuring of skin thickness. Sooner or later, most farmers will also require facilities for restraint to allow blood sampling.

Other less common veterinary procedures requiring restraint of hinds are assistance during the delivery of calves, insertion of intravaginal sponges or pessaries and the removal of hard antlers in those countries in which the removal of velvet is illegal.

Finally, there are the procedures connected with harvesting the crop. In those countries where it is permitted to remove velvet antlers, this will be an important part of the work of the handling yards. On all farms the ability to load deer into or out of lorries for transport is vital. At the time of writing, few farms slaughter deer in the yards, preferring either to shoot them in the field or transport them to abattoirs.

The Design of Yards

The 'Raceway'

It is generally accepted that deer farms are most easily worked if based on the principle of a series of fields, probably no larger than 4–6 hectares and possibly smaller, which interconnect by means of a 'raceway' (Fig. 5.3). This fenced corridor leads into the handling yards and allows deer to be driven out of any field on the farm into the yards at will. There is no requirement at all for the race to run in a straight line, except insofar as it is easier to fence. Once in the race the deer will move along with little stress at a walk or trot with only minimal pressure from behind.

Raceways vary in width from about 6 m to over 20 m depending on the size of the herd, and the need for space to manoeuvre vehicles. If too narrow, it may be difficult to move deer out of the fields and into the race and, if less than 15 m, it is almost impossible to move past a group of deer in the race without making a detour. A really wide raceway of 20 m allows grazing as a separate paddock.

The fencing for raceways is usually high-tensile deer netting, normally with the vertical stay wires at 150 mm rather than the 300 mm spacing. As the race approaches the yard it is usual to make the fences more visible and substantial by increasing the number of fence posts, attaching vertical wooden droppers or spacers, covering with plastic mesh, attaching wooden rails or replacing the wire fence with post and rail timber fencing. In all cases,

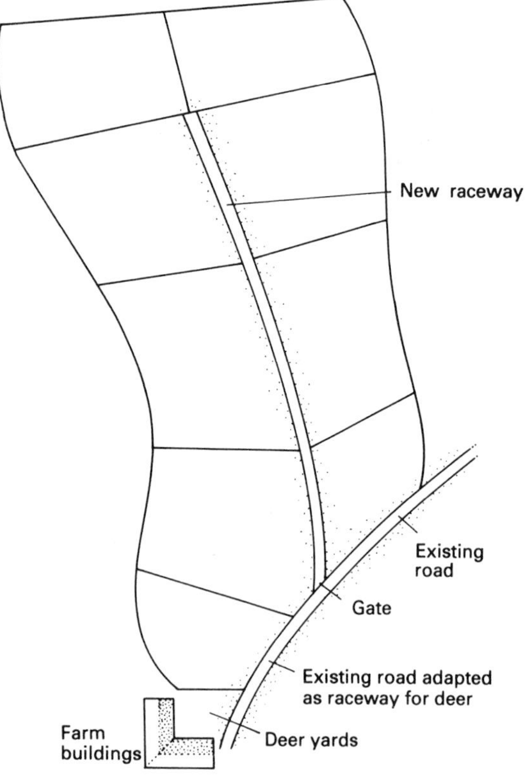

FIG. 5.3 Hypothetical layout of paddocks on a deer farm.

the fencing of the race must be at least 1.8 m high for red deer or 2.6 m increasing to 3 m as the yards are approached for fallow (Fig. 5.4).

Where there are gates in the raceway, there should be railed triangular exclusions to funnel deer past the constriction created by the gate posts (Fig. 5.5). These can often be advantageously planted with trees.

As the race enters the yard it is generally recommended that there is a bend — usually a complete right angle — so that deer moving along the race are not allowed to see that they are entering a cul-de-sac (Fig. 5.6).

The Holding Pens

The number of holding pens required depends on the size of the herd. These will generally receive deer from the raceway, hold them prior to handling, and again after handling. For large commercial herds it is usually necessary to use six or more pens. An 8 × 8 m pen will hold some 40 adult deer overnight or up to 80 deer for an hour or two.

The sides of the holding pens should be 2–2.5 m high. They can be made of solid plywood or complete close boarding, but it is usual now to use about 10 boards, measuring about 4000 × 150 × 25 mm, attached horizontally to round posts sawn in half lengthwise. These posts are usually at 1750 mm centres. The boards are nailed to give a 75 mm space between each. A similar effect can be created using vertical boards set with a 75 mm space. This open construction has two advantages, firstly, it is more economical and secondly, it allows the deer to see through to deer in other pens and to see people approaching. The effect is to make deer much more settled.

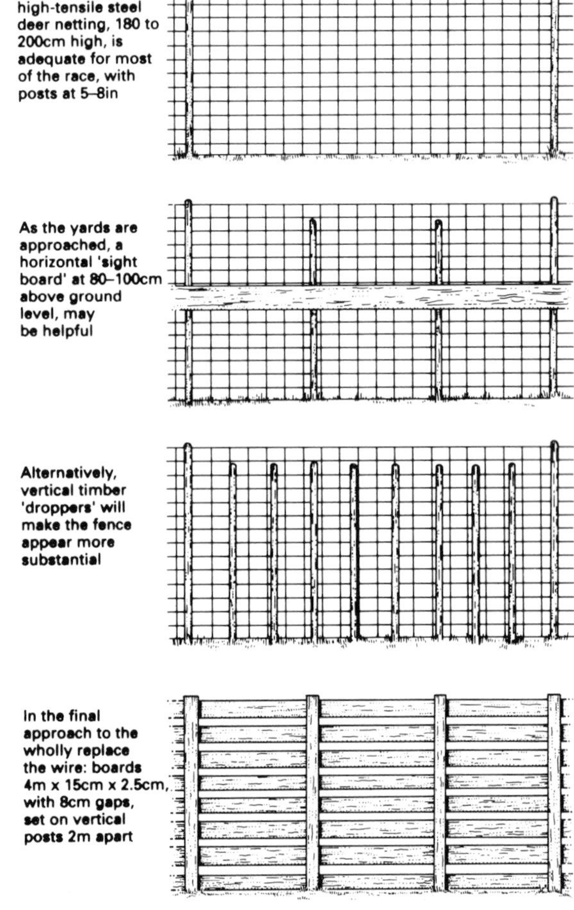

FIG. 5.4 Deer fencing in raceways becomes more solid as the yards are approached.

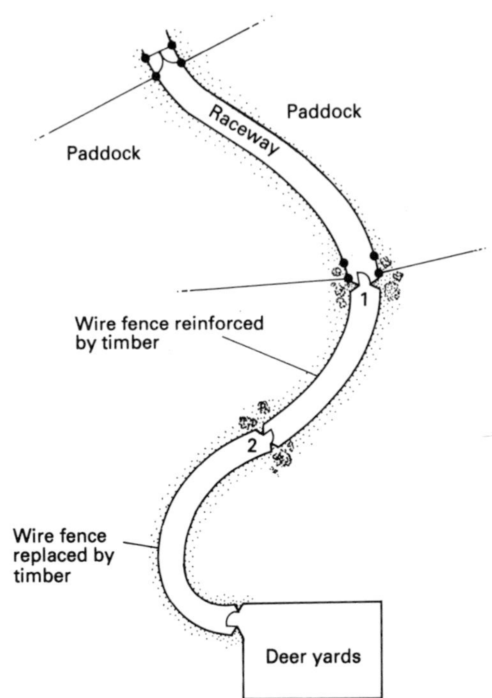

FIG. 5.5 Deer farm raceway designed to permit easy movement of deer into yards. ● indicates exit gates from paddocks. Gate 2 can be closed by pulling a wire at gate 1. Triangular exclusions, usually planted with trees, constrict the race at the gateways.

Gates from the holding pens need only be 1.2 m wide and, like all gates and fittings within deer yards, must be fitted with catches that are secure, easy to operate from both sides of the door and that do not protrude into the pen. Increasingly, catches are made simply of pins which slide into the gate or out of it into the post.

The Drafting Area

The design of the yards must permit deer to be brought simply and quickly from the holding pens into the working area and it is the design of this area about which there is most divergence of opinion among deer farmers. Many deer yards incorporate a circular drafting pen (Fig. 5.8). This has the advantage of having no protuberances so that deer can flow smoothly round. Many such circular yards have walkways on the top of the pen walls from which gates radiating from the centre can be closed to divide stock into convenient sized groups. The disadvantage of these pens is their expense and, where deer are being actually handled in the circle, the inconvenience of working in a triangular shaped sector. There is also the difficulty of making practically shaped pens adjoining the circle. Nevertheless, many modern yards still use a circular central drafting pen from which deer can be directed into holding and working pens.

Other yards which are extremely economical in construction, and are entirely adequate for the small and medium sized deer farm, are laid out in squares and rectangles (Fig. 5.9). The construction of the drafting area is very often of plywood but it may also be of sawn timber boards. In general, it is probably wise to keep the more frequently used working and drafting areas of solid-sided construction with the proviso that there should always be a means to allow people to see into neighbouring pens. This can be conveniently carried out by leaving a single space between boards at about 1.5 m above ground level.

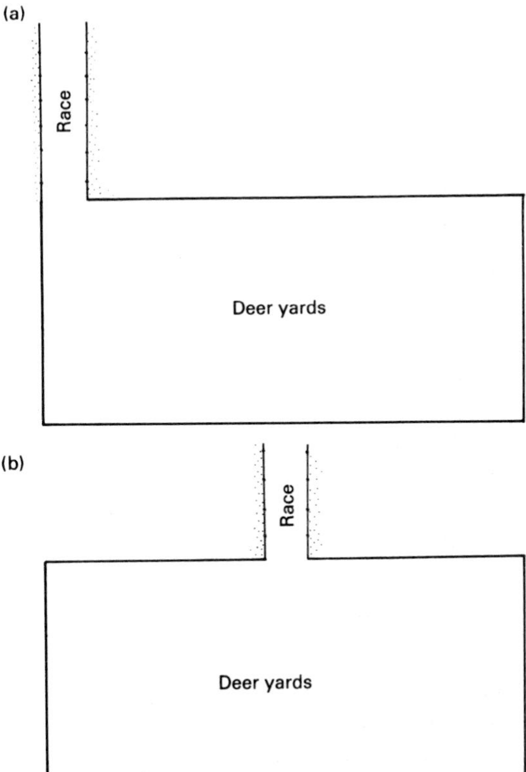

FIG. 5.6 It is a most important principle that deer should enter yards around a corner as in (a) *not* (b).

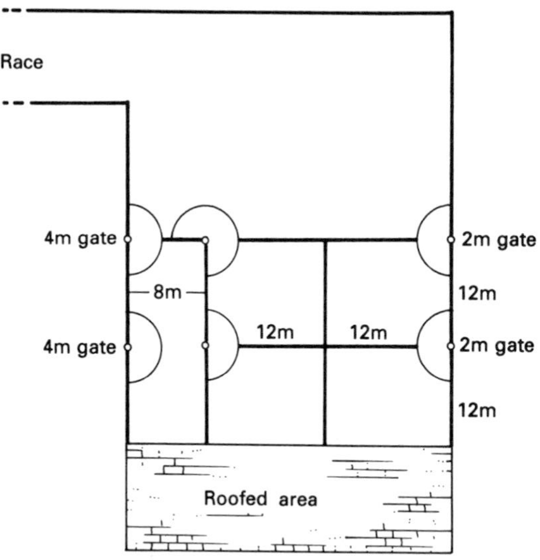

FIG. 5.7 Holding pens must be sufficient to accommodate comfortably all the deer required in any one muster.

The Working Area

Of the procedures listed earlier, drafting of deer into groups can clearly be carried out within the drafting area, but the regular procedures of inserting eartags and administering drugs are best done in a specific working area. This normally consists of a small pen measuring about 1.5 × 1.5 m or 2 m (Fig. 5.9). It must be solid-sided and must have good communications with an area restricted to human access only. In this way information can be passed through to someone who can write down eartag numbers or other notes. This pen will hold a small group of four or five hinds among which it is relatively easy to move.

In this area it is convenient to have a roof as low as 2 m, although this will preclude its use for large antlered stags. This low roof discourages hinds from rising up on their hind legs. There will probably also be a need for artificial light in the working area.

The Procedures

Weighing

It is quite realistic to use the entire working area as the floor of weighing scales. Modern electronic weighing systems allow one animal to be loaded and weighed, the scales then zeroed and the process repeated, to provide individual weights for each of a group of four or five animals.

Many farms have sophisticated computer-linked weighing systems permitting weights to be entered immediately on to an animal's record. The exit from the working area should allow animals to be released back into any of a number of holding pens so that deer can be drafted out depending on weight, udder development, etc.

Tuberculin Testing

Although TB testing and blood sampling can both be carried out in the working area, the very careful shaving of the neck, injection of the tuberculin and subsequent examination of the injection site which are required, particularly in those countries where avian tuberculosis is widespread, make it preferable to put deer into a 'crush' (Fig. 5.10). Blood sampling of large numbers of deer is also best done in a crush where handlers are less likely to have their feet bruised.

There are many different designs of deer crush but they can be divided into two fundamental types. Those in which the animal is restrained by its own weight and those in which hydraulic rams are used to sandwich the animal between padded cushions. In the first type, the deer are run along a short race and into a cage with a Y-shaped end elevation. Once in the crush, the floor is dropped by means of a lever and the animal's weight is taken by its thorax and abdomen with the legs no longer in contact with the ground. A handler is required to hold the deer's head. Once the procedure is completed a second lever permits the side of the crush to swing back, the animal sinks the few centimetres to the ground and can run out. More expensive and complicated crushes work on the same principle but with the sides swinging together under hydraulic pressure.

Removal of Antlers

Deer crushes are at their most useful for this procedure. Many yearling stags require their spikes to be removed before transportation to abattoirs and the use of a crush makes this procedure relatively safe and simple.

The removal of antlers from adult stags is less simple. The stag tends to back into the crush and

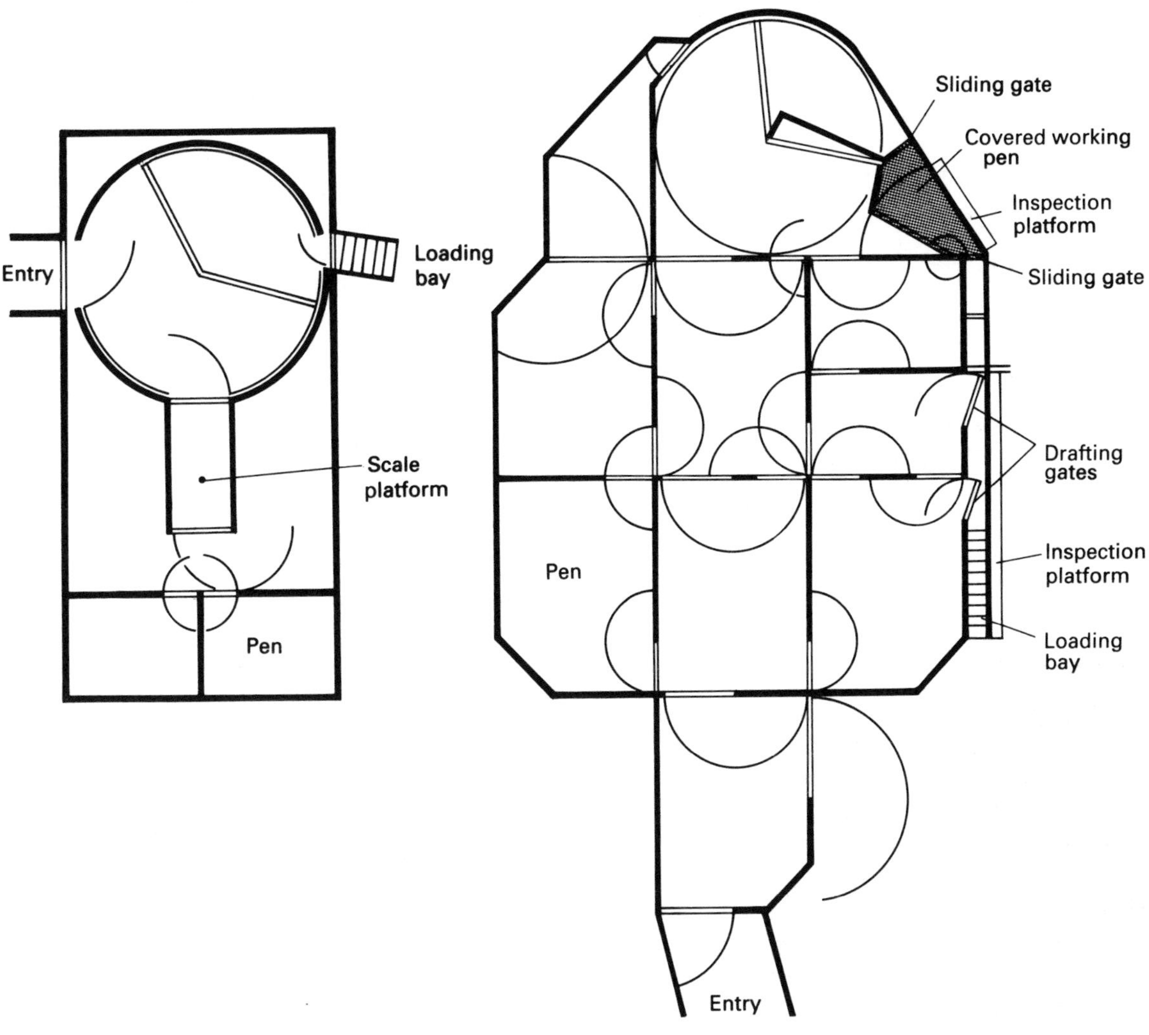

FIG. 5.8 Circular deer yards can be simple or complex.

remain there, stubborn and resolute. For this reason many deer farmers resort to the use of a 'dart gun' to tranquillize adult stags before they remove hard antlers.

In those countries where velvet antlers are removed routinely, the deer are usually sedated with xylazine and then restrained manually in the working area.

Loading Bays

Different countries have different loading systems for their livestock. In Britain, trucks invariably carry ramps so that yards require only a suitable sized gateway and a pen or race allowing deer to be driven up the ramp from a confined space. In New Zealand, animals are loaded up chutes into the lorry and the chutes are adjustable to the height of the lorry.

Use of the Yards

So far only the general principles of deer yard design have been discussed. The way in which the animals are handled is even more important. The finest set of yards in the world will not work with

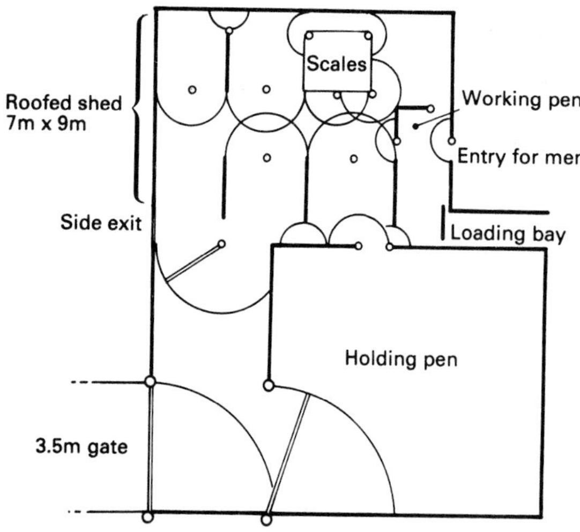

Fig. 5.9 Square yards may be as effective as circular pens and are likely to be more economical.

incompetent operators, while skilled deer handlers can manage in the most primitive buildings.

Before starting to handle deer, the yards must be checked to ensure that they are ready, the gates set correctly and the necessary equipment for the procedure is to hand. (Ideally the eartagging equipment, anthelminthics and other medicines should be kept in stores in the yards and there should certainly be an area into which deer have no access.) Unfamiliar objects should not be present — even a plastic bag blowing about in the wind can create problems.

The first, and often the most intractable, problem is moving the deer out of the fields into the raceway. Here, familiarizing the deer with the layout is crucial. If deer are turned directly from a lorry into a field, it can prove extremely difficult to move them through the gate on the first few occasions. Dogs may be useful but they must be well trained and responsive. It is usually worth leaving gates open overnight, as deer will explore much more boldly at this time.

Patience and two or three helpers may be required until the deer are used to being moved. It is also well worth using a feed bag to lead, as well as people behind to drive, the deer. The good stockman will spend enough time with the deer that they will come running to feed when called; deer learn this lesson very quickly given the opportunity. Food should always be distributed sufficiently widely for even the most cautious individual to get some food on each occasion.

Once in the raceway it may be helpful, if the race is particularly wide, to have two men stretch a length of hessian fabric between them as they walk. This eliminates any danger of an animal breaking back at the critical moment when the yards are entered. The hessian should always be kept tight and held above head height; this is a most effective technique for moving deer out of buildings or into lorries.

Some farmers use a long wire attached to a gate catch to allow them to close the gate behind the deer at the end of the raceway, but this is not necessary if the deer can move around a corner out of sight of the gate. Such techniques depend on the number of deer being handled at any given time and it is wise not to attempt to handle more than 100 deer on any one occasion, particularly where hinds are running with calves at foot or where antlered stags are present.

Once in the yards, deer should be given the chance to run through as many holding pens as possible, to allow them to settle out in small sized groups. Moving from holding pens into handling areas is generally very simple. By careful positioning, one man can usually cut individual deer out of a group quite easily by simply using his body or hands.

As with handling all animals, it is of tremendous importance to talk to deer when working with them. A constant flow of chatter means that the animals always know where you are and avoids the dangers of suddenly surprising them. They will respond by behaving much more quietly.

Fallow Deer

The foregoing pages relate primarily to the handling of red deer (*Cervus elaphus*) and their larger relatives, the elk or wapiti (*Cervus elaphus canadensis*). It is generally accepted that these temperate species of deer are the easiest to handle. If, however, red deer are not readily available or if there are fears of hybridization between introduced red deer and native wapiti, fallow deer farms have been established and systems have now been developed which enable even this more 'flighty' species to be handled efficiently.

So far as the raceways are concerned, fallow

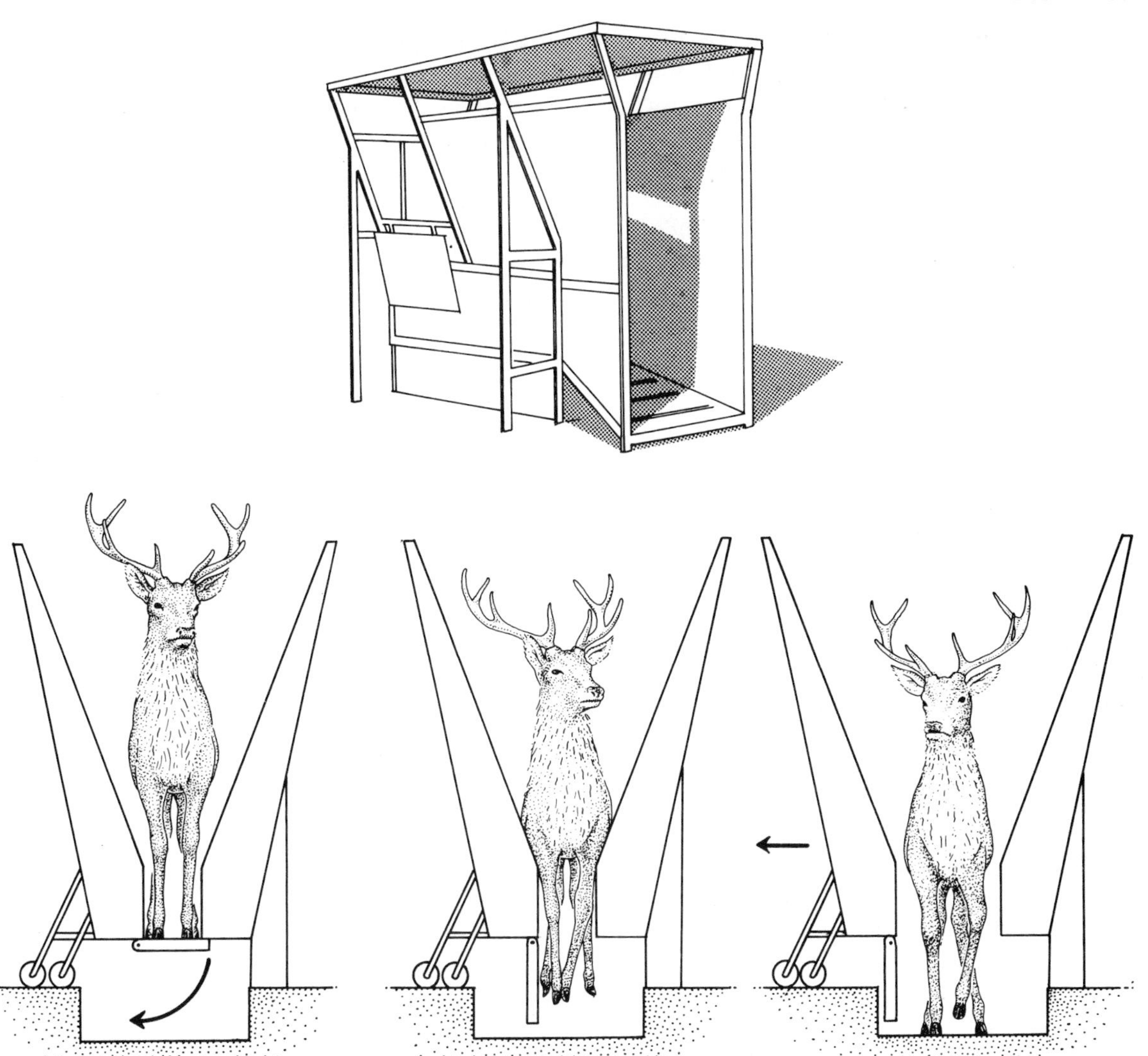

FIG. 5.10 A deer crush. Once deer have entered the common 'drop floor crush' the floor is swung away and the deer is restrained by its own weight. When the procedures are completed one wall can be swung back to permit the stag to walk out.

generally only require 3–4 m width although larger herds may require 15 m in the early part of the race. As with red deer, care in design at the point where the race enters the yards is important, with a gradual gradation in the solidity of the fence and a height increase up to 3 m. This height is desirable right through the yards for fallow.

The main point of difference in handling fallow deer is their response to light and dark. Once the deer are in holding pens, the lights should be dimmed to allow drafting into small pens. It is important that there are no extraneous points of light here. From the holding pens, the deer must pass into a light controlled pen where they are held with the lights on. When it is desired to move an animal forward into the handling area, the lights are dimmed and a door into a lighted tunnel opened. This normally encourages an animal to enter the tunnel, but it may be necessary to give further encouragement.

From the tunnel, deer normally enter a 'bale'. This is essentially a wooden box with a hole from which the animal's head and neck protrude allowing drenching, vaccination, blood sampling or TB testing to be performed. For procedures requiring access to other parts of the body, a crush, similar in principle to that required for red deer, is used.

Conclusion

No effort has been made to recommend any specific design of deer yards as it is a feature of deer farmers that each considers his yards the best in the world. This suggests that, provided the basic principles outlined here are adhered to, it probably doesn't make much difference what the precise layout is!

Further Reading

The Deer Farmer. Published periodical of New Zealand Deer Farmers' Association and AgPress Communications Ltd, P.O. Box 12-342, Wellington North, 16 Motorua Street, Thorndon, New Zealand.

Deer Farming. Published periodical of British Deer Farmers' Association, Editor Jessica Gould, 6 Maesygroes, Brechfa, Dyfed SA32 7RB.

Fletcher, T. J. (1986) In *Management and Diseases of Deer, a Handbook for the Veterinary Surgeon* (Editor T. L. Alexander). London, British Veterinary Association.

Yerex, D. and Spiers, I. (1987) *Modern Deer Farm Management*. Ampersand Publishing Associates Ltd, Box 176, Carterton, New Zealand.

6

Pigs

JOHN R. WALTON

Introduction

Handling and restraint of pigs relies to a great extent on treating pigs in a humane manner so as to prevent apprehension, fear and a need to escape. Several levels of restraint and handling are recognized. The first one attempts to mimic the actions of the owner by using simple techniques such as talking to or touching the pig, both of which can be combined with offering food to prevent the pig becoming agitated and disturbed in the presence of a stranger. The first level of restraint/handling relies upon establishing simple human–animal contact. The pig, like any other animal, will very soon recognise a person with a quiet authority developed from handling pigs and will usually respond to this in a very positive way.

Fig. 6.1 Voice is one of the best ways of making pigs aware of your presence.

Simple Human–Animal Contact

It is absolutely essential that, when approaching a pig housed on its own or in a group, it is quickly made aware of your presence. If this is not done, the pig will be startled and may injure itself or cause injury to others. This especially applies to recently farrowed sows with sucking piglets. Any sudden movement on the sow's part could very easily lead to a small piglet being stepped on or crushed. One of the best ways of making pigs aware of your presence is to use your voice (Fig. 6.1). Pigs will soon learn to recognize a voice, especially if that sound is also associated with the presentation of food. As with many other animals, pigs get used to a person's voice and will not be startled when that person appears. In some cases, sows may even be calmed and comforted by the sound of a known voice and the presence of a person who it recognizes.

A second very important aid to good husbandry is touch (Fig. 6.2). This applies especially to adult animals and those kept in confined stalls. Whilst

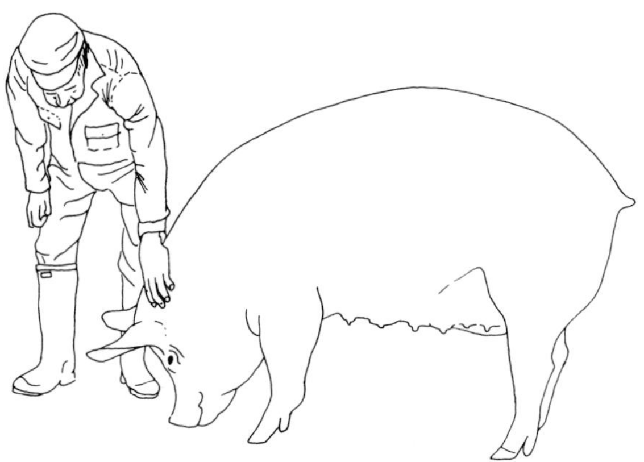

Fig. 6.2 Touch is a very important aid to good husbandry.

FIG. 6.3 The giving of food is one of the most effective forms of basic restraint in the pig.

touch may be less important to the infrequent visitor, it is of great value to the stockperson, farmer or owner who is in constant contact with his pigs. As with the voice, hand contact with pigs is often associated with feeding times and the pig will also become aware that a person is in the vicinity. It is not easy to quantify the value of using either voice or touch alone because on most occasions voice and touch are used together. Another benefit from using voice and touch is that people unfamiliar with pigs can gain a degree of confidence from this form of direct animal–human contact.

Probably one of the most effective forms of basic restraint is to provide the pig — especially an adult — with some type of food (Fig. 6.3). One of the primary instincts of the pig is to eat, and whilst it is eating the stockperson or veterinarian can perform a variety of simple tasks, as the pig will stay almost completely still. This basic method of restraint will permit the rectal temperature to be taken, the skin surface to be examined in detail, an injection to be given, the stage of pregnancy to be determined or simple wounds to be treated.

These three procedures — the use of voice, touch and food — will enable even an inexperienced stockperson to soothe and placate an apprehensive pig and to keep it still and in one place long enough for treatment to be carried out. It is essential, however, to remember that the tone of voice used should not be angry or high pitched, that a firm hand is applied to the pig's skin, and that the food that is offered is one that the pig is used to.

Use of Aids

The next level of restraint/handling relies upon using various aids to restrict a pig to a particular place, to help to move it from one place to another, or to prevent it going in an unwanted direction. These aids also provide a means of protection for the person who is attempting to administer some form of treatment.

The simplest form of aid is a small wooden board with a handle-hole near the top (Fig. 6.4). This board can be used to apply pressure to the side (Fig. 6.5), rear or front (Fig. 6.6) of the pig to indicate the direction you wish it to take. Excessive force should not be required to move a pig in any direction. However, pigs will refuse to move if the place you wish them to go to is dark, as would be the case when moving from the daylight to a very poorly lit building, or when asking the pig to climb a very steep ramp at the back of a vehicle, or to go along a passage that is too narrow or has too many sharp bends and turnings. The use of a small board can also be accompanied by giving food to the pig (Fig. 6.6) and using voice and touch (Fig. 6.7). On many occasions one may wish to use a board to indicate that forward movement of the pig is restricted with a solid object rather than the fragile open legs of the handler. A small board is very useful when walking alongside an adult pig to block off the route on the side that the handler is walking and to direct the pig to turn in the direction away from the board.

Probably one of the best uses of the small board is when it is held by the stockperson or owner at the side of the pig's head to prevent the pig from biting or head butting the veterinarian when he is giving

FIG. 6.4 A small wooden board is the simplest form of aid to pig handling.

FIG. 6.5 The small board used to apply pressure to the side of a pig.

FIG. 6.7 The small board can also be accompanied by the use of voice, touch or the giving of food.

an injection. Very often an adult pig can be kept still if food is given and a small board held against one side of the head and slight pressure applied to keep the pig close up to an adjacent wall. This procedure is very effective when carrying out simple manipulations of the rectum and vaginal examinations. Care must be taken to ensure that the small board is not used to administer a blow to the pig or to injure it in any way. The cardinal rule to remember when attempting to handle or restrain any adult pig is that pure physical force is usually counter productive and can be very dangerous for the handler. This is especially true when one remembers that an adult sow or boar can weigh up to 300 kg (700 lb). When moving adult pigs, or when controlling them prior to an examination, a firm grasp of the tail with a little forward pressure will help to direct the pig along a desired pathway or keep it in a particular position (Fig. 6.8).

A variation on the small board as a handling device is a very much larger board, often used by

FIG. 6.6 The small board used to apply pressure to the rear or front of a pig.

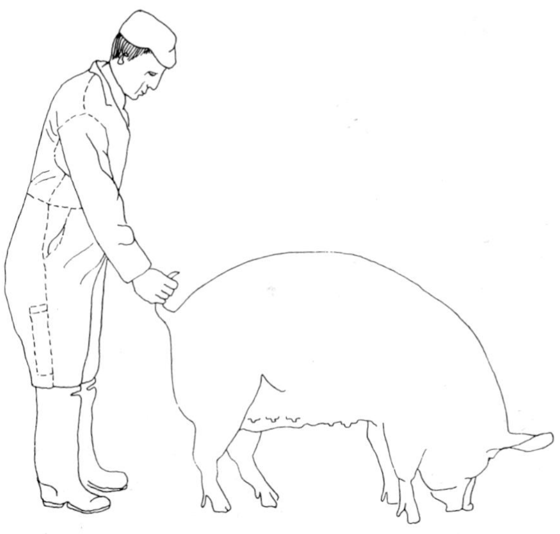

FIG. 6.8 Forward pressure on the tail is very useful when directing pigs along a desired pathway.

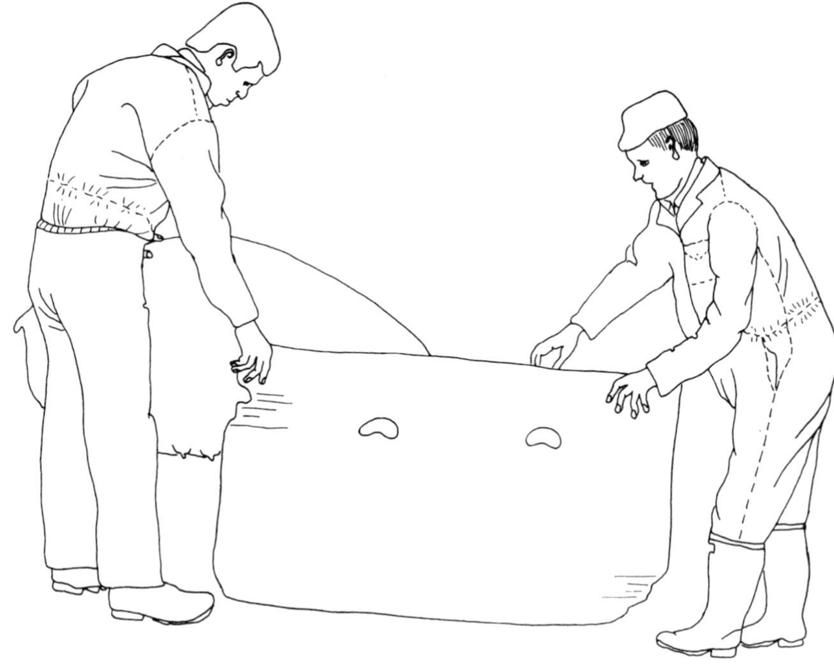

FIG. 6.9 A large board is often used by two people to restrain adult pigs.

two people (Fig. 6.9). This larger board is very useful for restraining a single, large pig or several smaller pigs in a corner of the pen. Once again, this larger board provides much greater protection for the veterinarian attempting to carry out a painful procedure, especially on a large adult pig (Fig. 6.10). It is still possible to use voice and touch whilst applying side pressure with a large board. The skilled stockman will continuously communicate with his pigs while he is in their presence. Some will talk continuously; others will whistle or sing. When moving pigs with the aid of a board it must be absolutely clear to the pig the direction in which it is being asked to move. The passageway should be well lit and there should be no sharp projections or damaged walls that could cause injury to the pig.

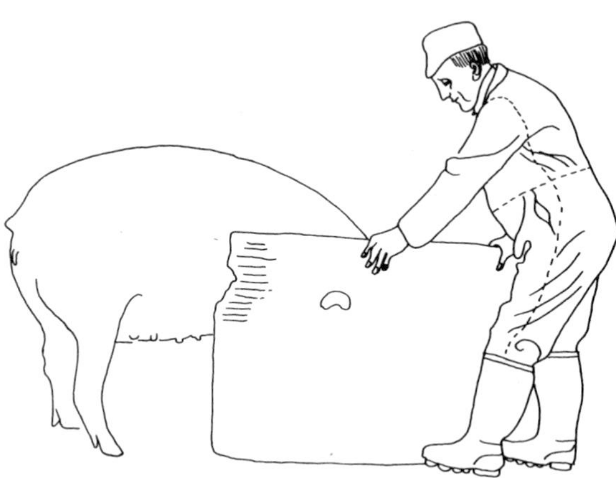

FIG. 6.10 The large board also provides good protection for the veterinarian.

General Handling

The next level of restraint relies on positioning the pig to facilitate the application of certain surgical or other procedures. These procedures include castration, tail docking, tattooing, teeth clipping, and iron injections. Many of these procedures are performed on very young piglets that are small enough to be picked up and held in the operator's hands or arms. The very small piglet can be picked up and held firmly with both hands across the operator's chest (Fig. 6.11). This position will enable a variety of simple techniques to be per-

FIG. 6.11 Small piglets can be picked up and held against the operator's chest.

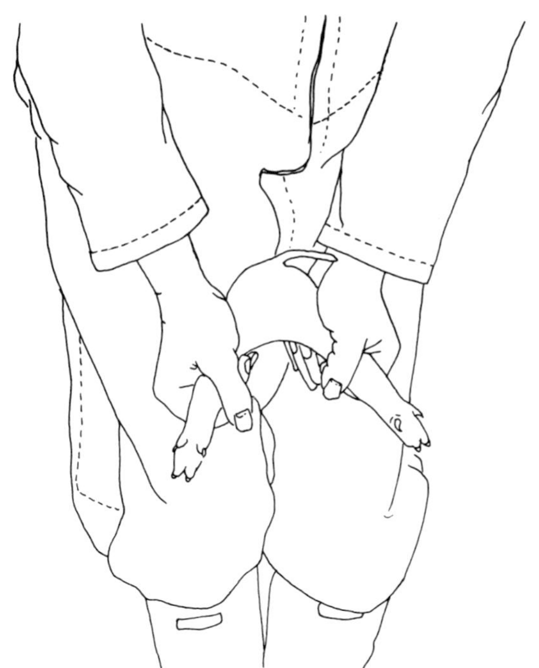

FIG. 6.12 Piglets can also be held by the hocks with the fore-quarters being restrained by the stockman's legs.

formed, such as taking the rectal temperature, administering an injection and taking a skin scraping. When picking a piglet up from the floor it should never be picked up by the ears. This is a very painful act and may even lead to the ear being injured. Young piglets should always be caught and picked up by grasping a hind leg, or legs, and then holding them as in Figs. 6.11 and 6.14. When holding a young piglet as shown in Fig. 6.11, it is essential that the piglet is held firmly so that it feels secure and safe. In this way it will keep still and generally will not struggle, as it would if the operator's grip was weak and insecure. Very small piglets can also be carried and moved from place to place in a small box on wheels or a high-sided wheelbarrow.

Castration

Several methods of restraint are used for this surgical procedure. All of them rely upon the piglet being held securely whilst presenting the scrotal region to the stockperson. The first method shows the piglet's fore quarters being held between the stockperson's legs whilst the two hind legs are held firmly around the region of the hocks (Fig. 6.12). This method keeps the piglet in an upright position and is said to avoid undue disturbance. A variation on this method involves the stockperson holding two legs (a front and a back) in each hand (Fig. 6.13). This, however, does result in the piglet being held in a head-down position with no support to its upper body at all. A third method is simply to hold the piglet by the hind legs, head down with the back of the piglet against the stockperson's body (Fig. 6.14). All these three methods have their advantages and disadvantages and the method finally used will rely upon personal preference. Many variations of these methods are used and various metal or wooden frames have been produced so that the whole of the operation of castration, tail docking and iron injections can be carried out by one person.

Tail Docking

This procedure is carried out using essentially the same holding positions as for castration except that there should be provision to ensure that the tail is measured before being cut.

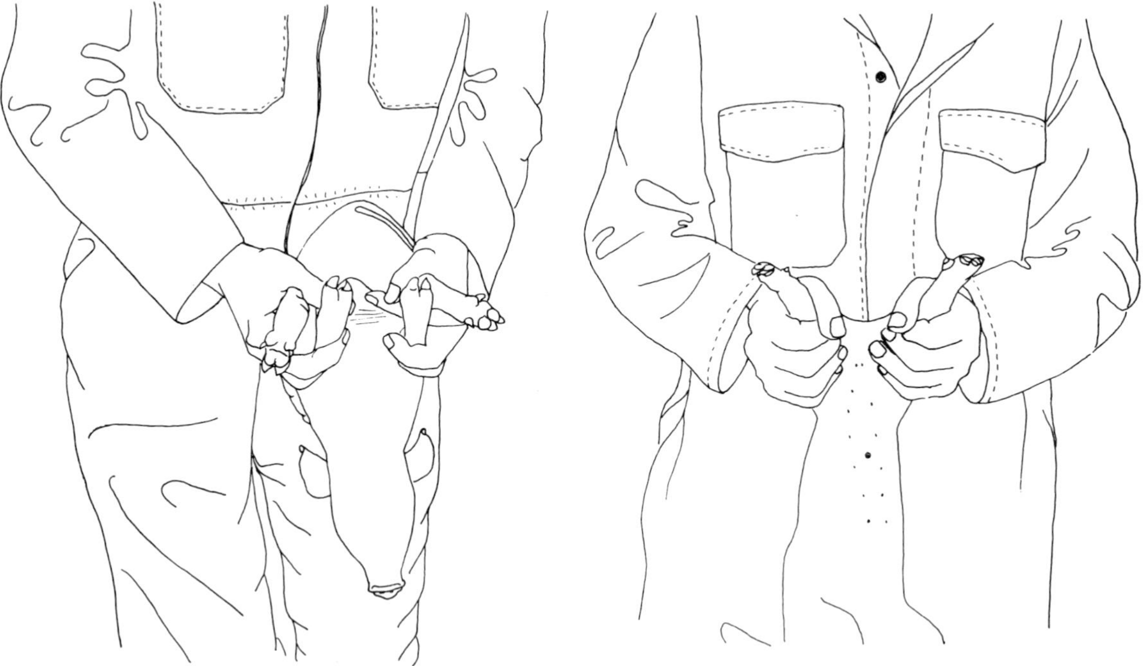

FIG. 6.13 A piglet being restrained by the stockman holding two legs in each hand.

FIG. 6.14 A piglet being held by the hind legs with the back against the stockman's body.

Ear Tattooing and Teeth Clipping

These two procedures are generally carried out with the piglet being held in the stockman's arms with the head uppermost (Fig. 6.11). The piglet is held firmly against the stockman's chest. This appears to comfort the piglet and so avoids undue squealing. If the procedure is being carried out single-handed, the piglet can either be held between the stockman's legs or on a special metal cradle that can be attached to the back of a feeding trolley.

Immobilizing the Larger Pig

Restraint of the Bigger Pig to Facilitate Bleeding, Intravenous Injections or Simple Surgical Manipulations

Wire or rope snares (loops)

The most useful method to restrain a large pig for these procedures is to use a small wire or rope snare that has a running loop at one end. The loop is slipped over the upper jaw and positioned behind the tusks so that when it is tightened by pulling back on the handle, the pig's head is held in a forward and upward direction (Fig. 6.15). Because the loop is positioned behind the upper tusks it cannot slip off if the pig pulls vigorously or shakes its head. Generally speaking, once the loop is around the snout, the pig will pull backward, and very rarely will it move forwards (Fig. 6.16). If blood has to be withdrawn from a boar, it is preferable to use a spade snare (Fig. 6.17) as this will give the stockperson better control over the pig's head and a more substantial handle to hold.

Holding or restraining tongs

Tongs (Fig. 6.18) are now infrequently used as they are very inferior to the wire loop. If they are to be used they are best used on smaller pigs and the snout grasped by the jaws of the tongs in the same region as with the wire loop, i.e. behind the upper tusks.

FIG. 6.15 Use of the small wire snare in older pigs.

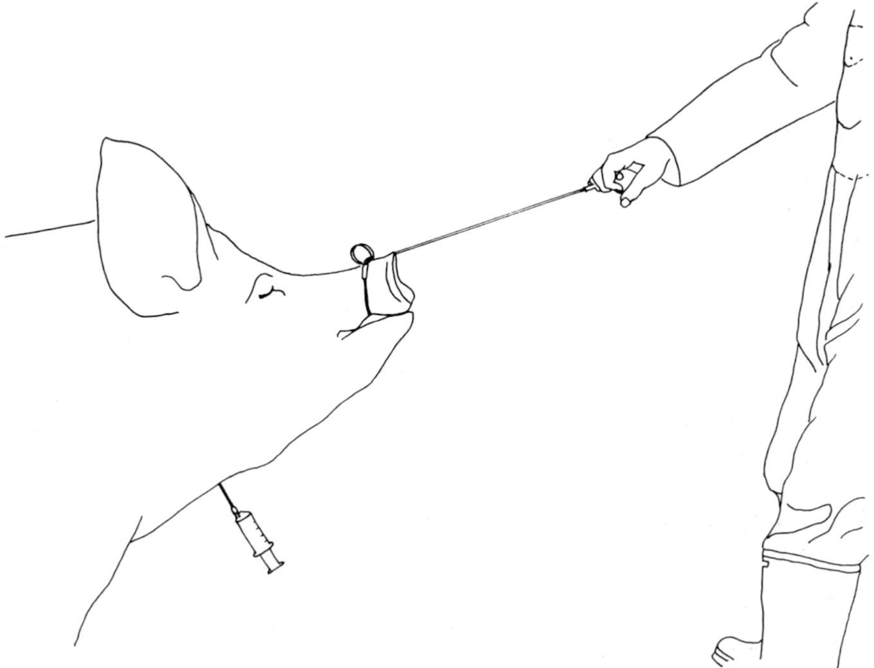

FIG. 6.16 A pig will generally pull backward when the loop is attached.

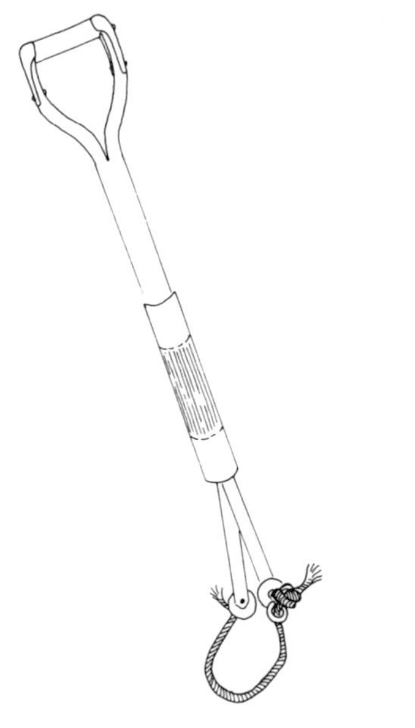

FIG. 6.17 A snare is best used when controlling boars.

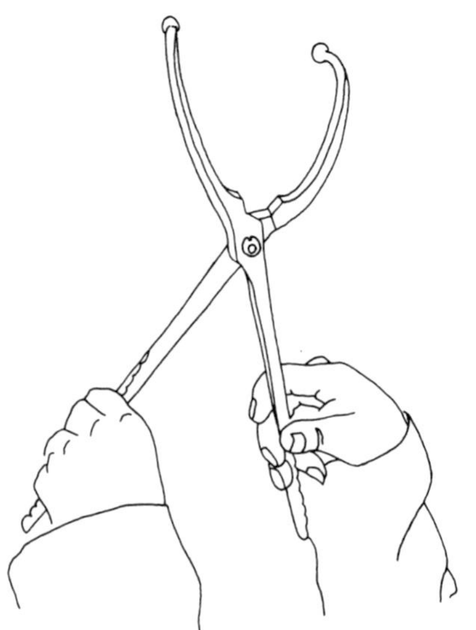

FIG. 6.18 Holding or restraining tongs have been used to control smaller pigs.

Chemical Restraint

Sedation

When fractious adult pigs have to be handled or when sows are savaging their newly born piglets, chemical sedatives can be injected and these will generally sedate the pig and enable any handling to proceed safely.

Anaesthesia

This procedure can only be applied by a veterinary surgeon as it requires special anaesthetics to be administered into a vein. Anaesthetics can also be applied in gaseous form once the pig has been premedicated to enable a gas delivery tube to be inserted into the trachea through the mouth. This procedure is used for surgical operations or when painful manipulations have to be made to the feet or limbs.

Final Observations

The successful handling and restraint of pigs requires the stockman or handler to be quiet, efficient and purposeful. On no account must excessive force be used or cruel procedures applied, such as electric goads or wooden sticks. The failure to handle pigs properly will result in animals developing a fear of the stockman or becoming ill-tempered or even being injured. Severe stress may also be produced by irresponsible handling to a point when the pig may actually die. The sign of a good stockman is one who understands pigs, doesn't bully them, and ensures that any movement is made using quiet persuasion and the least force.

One final point is that one must always remember that pigs are not always quiet and docile. Some can be very bad tempered; others easily frightened. Under these conditions the pig may rush at and try to bite the human intruder. This is especially the case with sows or gilts that have recently farrowed and are trying to protect their new-born piglets. Special consideration should always be given when handling boars. These are often very large and heavy with two pairs of very sharp protruding tusks. Severe injury can be inflicted upon strangers, such as veterinarians giving injections and also upon careless stockmen. It is important, therefore, always to carry a board for protection against the boar's teeth and to also help to restrain the boar during handling. If at all possible a second person should always be present when working boars are being handled or restrained.

Further Reading

English, P. R., Fowler, V. R., Baxter, S. and Smith, W. J. (1988) *The Growing and Finishing Pig*. Farming Press, Ipswich.

King, J. O. L. (1978) *An Introduction to Animal Husbandry*. Blackwell, Oxford.

Sainsbury, D. (1986) *Farm Animal Welfare: Cattle, Pigs and Poultry*. Collins, London.

7

Chickens, Turkeys, Ducks and Geese

PETER E. CURTIS

Introduction

The term poultry, which is derived from the old French term poulterie, is used to describe collectively the domesticated birds of the farm yard, such as chickens, turkeys, ducks and geese. It is sometimes extended to include other species, such as quail and game birds, which are not discussed here. This extension of the term is used particularly in disease legislation.

There is a popular idea in some quarters that chickens, turkeys, ducks and geese, being less 'advanced' than mammalian species, are simple, unintelligent animals, with little awareness of their environment, and that they can accordingly be treated as 'second class' animals. The term 'bird brained' is often used of human beings who are held in mild contempt, but in fact we are curiously ignorant of the level of awareness of poultry and there is now a view that they should be treated more kindly than in the past. Clearly the original species from which our domestic birds were derived were highly successful and resourceful animals, capable of surviving in and adapting to different environments. Our modern commercial variants still retain much of their original genetic potential, notwithstanding the fact that they have been intensively selected for characteristics which enhance their productivity. Good stockmen always treat their animals with understanding and they have an empathy with their charges which contributes to the success of the farm enterprise and to the animals' welfare.

Chickens are a domestic form of *Gallus gallus*, the Red Jungle Fowl, and turkeys, ducks and geese are derived respectively from *Meleagris gallopavo*, *Anas platyrhynchos*, and *Anser anser*. All these species are birds and therefore have some common factors including beaks, feathers, wings, and the specialized avian respiratory system, involving air sacs, with diverticula extending into bony structures.

Vision

Like all birds, poultry have good all round vision which is helpful in detecting approaching predators. Their acuity of vision enables them to peck, very precisely in the case of chickens, at interesting food items on the ground. Reducing the birds' ability to see can thus be an effective way of reducing their general awareness when it is necessary to catch them for transport or for other purposes, such as weighing. Poultry are also influenced by novel clothing worn by visitors and also by the appearance of any equipment being used. Bright clothes or highly reflective equipment can cause alarm and, in some cases, mortality due to smothering.

Hearing

The auditory sensitivity of poultry is impressive and much of their time is spent exchanging sound messages which may be presumed to provide some mutual reassurance of safety. When strangers appear, an alarm call may be emitted which may, in extreme cases, cause serious alarm throughout a large flock and an extensive flock reaction. For this reason visitors to poultry flocks should approach with caution and indicate their arrival by tapping on the entrance door of the poultry building before entering.

Catching

When catching birds a smooth style of movement is essential, as any sudden movements or lunges will cause the birds to take avoiding action and become almost impossible to catch. Such episodes can result in birds being packed into corners and smothered, others being injured when fleeing and egg production suffering. Even experienced persons can cause alarm in a flock through some quite slight, but unexpected, movement. In very small flocks, where there may be a great deal of space for the birds to take evasive action, accidental injury to individual birds is an increased risk. It may be advisable to invite farm staff to capture any birds which have to be examined, at least initially, since the regular staff will be less alarming to the birds than the unfamiliar visitors. Birds will, however, soon adjust to strangers.

Tonic Immobility

Tonic immobility is the condition in which some birds remain immobilized when placed on the ground after a period of being handled or transported. It has been particularly noted in broiler chickens after transport, but can also be detected in other chickens. It seems likely that it is a fear response which, in the wild and aided by camouflage, helped birds to escape detection by predators.

Diversity Amongst the Species

There is a great diversity among chickens and other poultry as a result of years of natural breeding plus generations of selective breeding to achieve a variety of targets from the ornamental bantam to the fighting cock, and from the high egg producer to the rapidly growing poussin. This diversity, so apparent in physical appearance, also extends to behaviour.

Chickens

Chicks

The chick is extraordinarily resilient. Having hatched from the egg after exercising its lungs for some hours in the shell, it is ready to explore the world and learn how to find food. Newly-hatched chicks are equipped with insulating down feathers, and an internal supply of nutrients in the form of a yolk sac which helps them to survive the first three days of life.

Hatchery staff are trained to eliminate any chicks with moist navels, which might facilitate yolk sac infection, and to exclude any chick with obvious deformities such as crossed beaks or splay legs. Chicks are easily handled and can readily be moved by conveyor systems to locations where they may have to be handled again for such purposes as inocculation, trimming of the beak tip, if necessary, and sexing. The combs of male chicks (up to 72 hours) may be removed (or dubbed) with scissors and, in the case of turkeys, the snood may be removed.

After counting, chicks are placed in rigid boxes or crates which are loaded into special air-conditioned and temperature-controlled vehicles for transport to the farm.

Newly hatched poults, ducklings and goslings have a similar introduction to life, but they do not receive Mareks vaccine, and only turkey poults might require beak trimming.

Older Chickens

Catching

Older chickens may be caught by grasping the body with two hands with the effect of holding the wings in place and preventing flapping (Fig. 7.1). This ideal approach is not always possible and one may have to resort to catching by means of the legs, which can both be grasped with one hand, the other hand being used to support the bird from below in the upright position. The bird can then be held with the legs grasped and the body resting on the hand (Fig. 7.2) to allow it to be handled for condition scoring, general examination, show judging, or assessment of laying status by measuring the gap between the pelvic bones with the fingers (Fig. 7.3).

Only by handling a representative number of birds in the flock can the stockman assess the general bodily condition and health of the flock. Simply observing from a distance does not reveal what lies below the feathers and thin or overfat birds may appear superficially normal. Regular weighing is desirable (Fig. 7.4). Weighing of groups may be performed with the birds restrained in a crate.

FIG. 7.1 Holding the wings of an adult hen to the body to prevent flapping.

FIG. 7.2 A show bantam held for examination.

If the chicken to be caught is at large, a fisherman's keep net may be used to catch the bird so that it can be held.

Holding and carrying

When a chicken is caught, it must be held firmly as this helps to reduce wing flapping and struggling

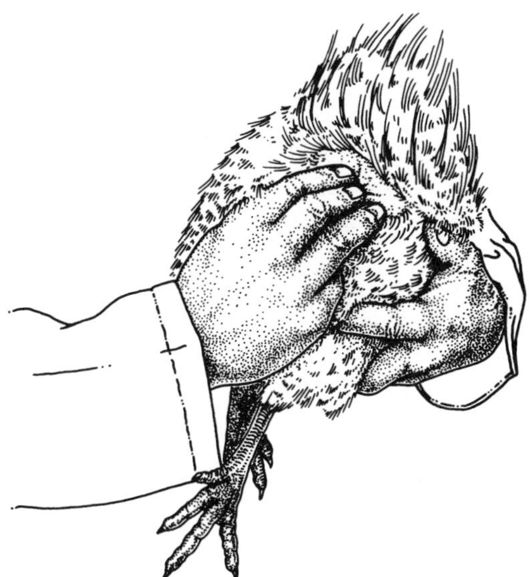

FIG. 7.3 Checking if a hen is in lay.

FIG. 7.4 Weighing a live bird with a spring balance.

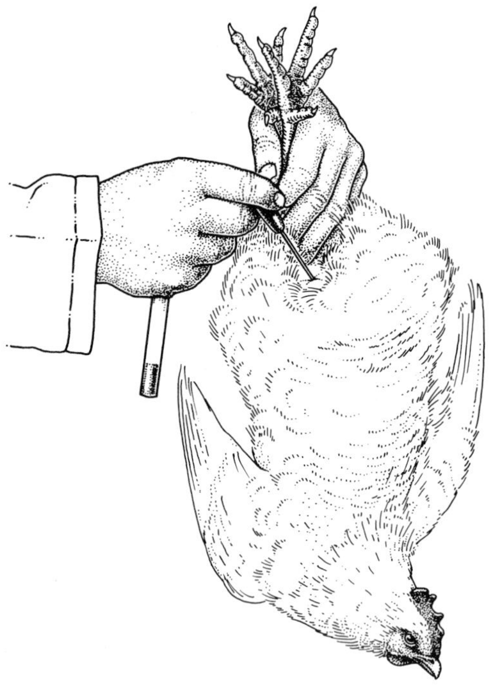

FIG. 7.5 Single-handed swabbing from the cloaca.

which could result in damage to the bird itself and would certainly provoke alarm calls by the birds in the vicinity and widespread alarm in the flock. Usually a chicken will struggle for a short period when caught and then relax for a period before once again attempting to struggle free. Thus firm but responsive handling will be most successful in keeping the bird comfortable, whereas loose holding of the bird may allow it to free one or more limbs with the risk of injury.

In some circumstances it is necessary to carry adult chickens by the legs, and two legs must always be held or injury to leg tissues will result. Leg carriage is often used when weighing birds and when it is not feasible to use the upright style. (It is not permitted to carry chickens, or any poultry, in a public auction market in the United Kingdom using only the legs and with the head downwards.) If chickens suffer damage to the muscles, tendons, or ligaments of one leg, as a result of careless handling, they will subsequently be severely handicapped and may have difficulty reaching food and water in some systems. They will also be prone to attack by other birds.

Evidence has accumulated that hens which have been kept in batteries are much more prone to bone fractures than birds on deep litter or free range, owing to increased skeletal fragility. It is, therefore, particularly important that battery hens are handled with special care to avoid fracturing wings, pelvis, legs, sternum, or other bones.

Swabbing and sampling

When swabbing from the cloaca, it is convenient to hold birds by their legs with head downwards if working single-handed (Fig. 7.5). With oral, oesophageal, or tracheal swabbing, assistance is necessary to hold the bird in the upright position (Fig. 7.6). When blood sampling from the jugular, it is customary to use the larger right jugular, and an assistant is required to raise the vein and hold the bird (Fig. 7.7). Otherwise the wing (brachial) vein may be used.

FIG. 7.6 Swabbing from the oesophagus.

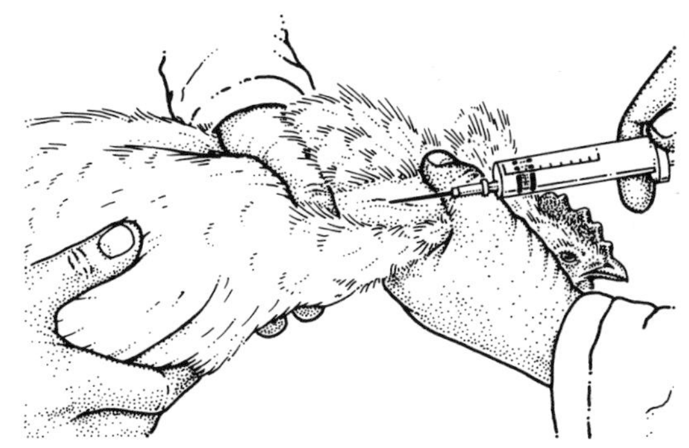

FIG. 7.7 Bleeding a hen from the right jugular vein.

Broiler Chickens

Some 600 million broiler chickens are reared in Britain each year and these strains of chicken have been selected for rapid growth, a quality which means that the farmer will handle some six crops per year per house. As these chickens never reach true physical maturity in most systems, their limbs are not fully calcified and therefore vulnerable to injury. It is, however, necessary to convey them from the intensive house to the slaughter plant with minimal stress.

Catching

If working during the hours of daylight, the sheds should be darkened by draping dark plastic sheets over any light inlets. Catching at night requires only that lights be dimmed as much as possible. The crates are then brought into the shed, usually in the form of a module conveyed by a fork lift truck. The modules comprise some 16 crates, each of which is 255 mm high and 910 mm square. The number of birds they can hold is assessed in relation to the weight of the bird, the distance to be travelled and the climatic conditions anticipated for the period between loading and unloading. Fairly close packing will prevent birds being thrown about when in transit on the road, but if too many birds are put in a crate, additional stress and even suffocation may ensue. Sheeting may be employed externally in cold weather or in heavy rain and drivers should always drive with a proper respect for their livestock passengers.

Catching of broilers is a labour-intensive operation involving skill, strength and considerable agility. The task has to be performed at speed to meet the deadlines of the slaughterhouse, but damage must be avoided. It is customary to grasp birds by both legs (Fig. 7.8) and to use both hands to raise two groups of birds, which must then be placed gently, but quickly, in the crate. Broilers must not be caught by the wings as their bones are easily damaged and there will be fractures or dislocations which will cause suffering to the bird and down-grading at the slaughterhouse. This damage is readily detected after slaughter, when it will appear as bruising with local haemorrhage. Catching teams must be notified if any evidence of catching damage is detected at slaughter, so that their techniques can be corrected. A financial penalty may also be imposed.

If fork-lift trucks and modular crates are not employed, it may be necessary to carry birds for a considerable distance to the end of the shed with the increased risk of damage to their legs. Passing birds from hand to hand is also likely to increase the risk of injury as a result of the inevitable fumbling which will occasionally occur.

Mechanical 'harvesting' machines have been devised to collect broilers from the floor and convey them gently to crates, and broiler chickens, being somewhat unresponsive, can accept these systems without significant stress. Suitable houses with minimal roof support columns are essential for the machines to work satisfactorily.

Individual broilers for clinical examination or culling purposes should be picked up bodily (Fig. 7.9) rather than by the legs and should be held with their heads up, and wings close to the body, to prevent wing flapping and possible wing injury. They should be returned to the floor gently and not simply dropped.

Turkeys

Adult turkeys are regularly handled for breeding purposes, since artificial insemination is widely practised and semen has to be collected from males, and hens caught for insemination. Other turkeys may be handled for various management purposes such as veterinary examination, culling, transport, vaccination, and weighing. Younger turkeys are not handled as frequently, but may still

FIG. 7.8 Carrying broilers aged 47 days.

FIG. 7.10 Catching a turkey by the right wing and left leg.

FIG. 7.9 Carrying an individual broiler.

FIG. 7.11 Lifting a turkey using a wing and the opposing leg.

be caught for administration of vaccines or drugs by injection.

Turkeys are driven into a catching pen, or reduced area, to be caught by experienced staff. The technique is initially to catch the individual bird by one leg and the opposite wing (Figs. 7.10 and 7.11), then to grasp both legs in one hand and place the bird on its breast on a bale of straw (Fig. 7.12) or similar soft object, or on the catcher's knee. Such catching requires a confident approach and good coordination by the handler. Talking to the birds prior to catching is helpful and the handler may whistle to ensure the turkey's head is upright at the time of catching. Once held in the described position, the turkey can be manipulated for clinical examination, swabbing of the mouth, oesophagus, trachea, cloaca or other site, the administration of drugs by injection or orally, or collection of blood from the wing vein (Fig. 7.13).

Careful catching and handling is essential if bruising, or even fractures and dislocations, are to be avoided. Turkeys may achieve considerable weights (Fig. 7.14) and strong catchers are required.

Ducks

Catching and Holding

Ducks are social animals with a group lifestyle that allows them to be driven, with suitable soothing words, from one place to another. Thus, for catching purposes, the ducks can be driven to a catching area close to the crates in which they are to be placed and all the ducks will move in a cohesive mass to the selected area, their heads raised and eyes watching for signs and signals. Ducks can see quite well in dark conditions as they are night feeders in nature, so dimming the lights is not helpful for catching purposes.

The duck has no true crop but there is a widening of the oesophagus which is used for holding food. Pressure on this area must be avoided or the food will be regurgitated. It is advisable to avoid handling the soft tissues in front of the neck. The trachea is, however, relatively sturdy compared with the chicken.

Ducks may be caught by the head, being grasped from behind in a form of scooping motion with the palm behind the head and the thumb and forefingers on either side of the neck. The front part of the neck is left free of pressure (Fig. 7.15). The duck can then be gently moved forwards and upwards and the wings grasped (Fig. 7.15). The neck can then be released and the duck supported from beneath the body. The wings may be grasped with one hand and the other used to provide support below the body (Fig. 7.16).

Pressure must not be applied to either the rib cage or the breast, as damage to skeletal or internal structures can occur, especially in young ducks. The duck must not be held by grasping the chest or the abdominal area. Similarly the legs should not be used for catching or holding the duck, as damage is likely to result.

After handling a duck it should be placed on the ground with care, preferably with a slight forward motion and the feet in the correct position for a safe 'landing' so that no jarring effect is experienced by the duck. Ducks should not be simply dropped to the ground as injury would probably occur.

Blood Sampling

Blood sampling can be performed using the wing vein with the duck sitting and the wing and vein raised (Fig. 7.17), or using the leg vein on the anterior aspect of the hock area with the duck recumbent on its back (Fig. 7.18).

Geese

Adult geese vary in size according to the breed or strain. Some are little bigger than a duck, whereas others weigh over 10 kg. Geese are considered to be perceptive and intelligent animals and should be treated as such.

Catching and Holding

Several methods for catching geese may be considered. If it is possible to drive them into a small enclosure in small numbers, they can be picked up individually by first grasping the goose by the base of the wings (Fig. 7.19). It is sometimes the custom to catch the goose as it is attempting to depart from the small enclosure by a narrow exit specially provided for the purpose, the goose having a natural tendency to wish to escape from such

FIG. 7.12 Turkey held with sternum resting on a straw bale.

FIG. 7.13 Turkey held with a wing raised for blood sampling from the wing vein.

FIG. 7.14 Holding a turkey for weighing.

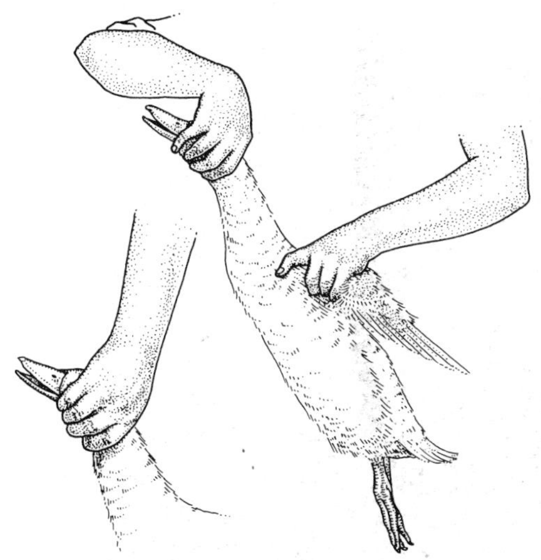

FIG. 7.15 (*left*) Catching a duck by the neck. (*right*) Raising the duck by wings and neck.

FIG. 7.16 Carrying a duck on the hand, restraining the wings.

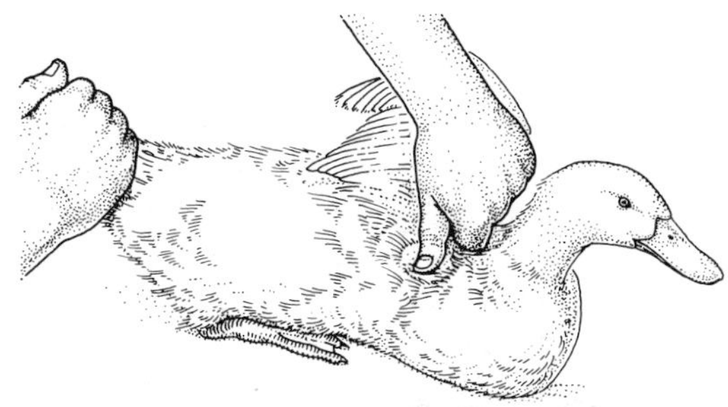

FIG. 7.17 Duck restrained for blood sampling from the wing vein.

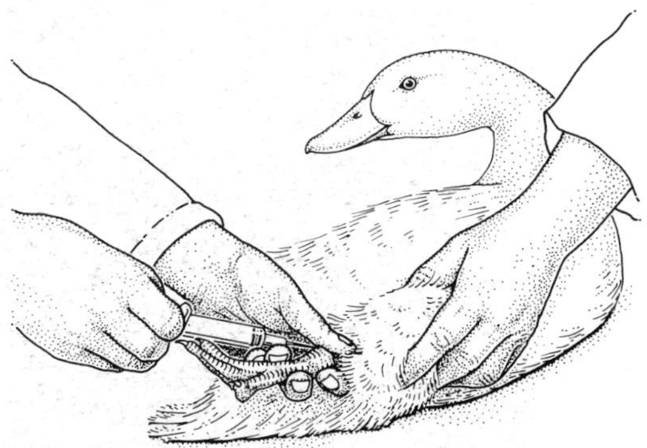

FIG. 7.18 A duck held for bleeding from the leg vein.

FIG. 7.19 Grasping both wings of a goose.

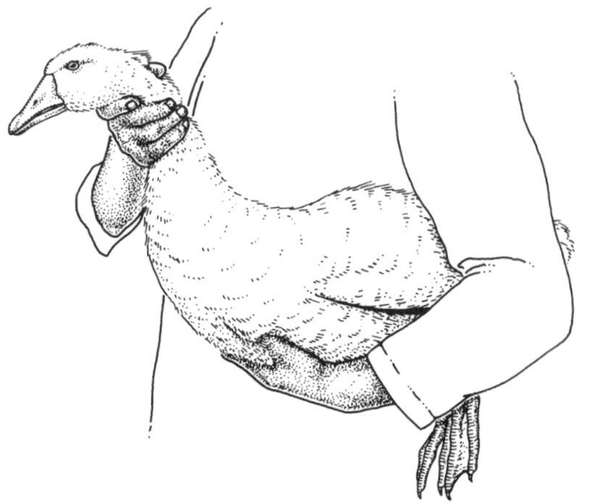

FIG. 7.21 Holding a goose head forward.

FIG. 7.20 Raising a goose to the under-arm-head forward position.

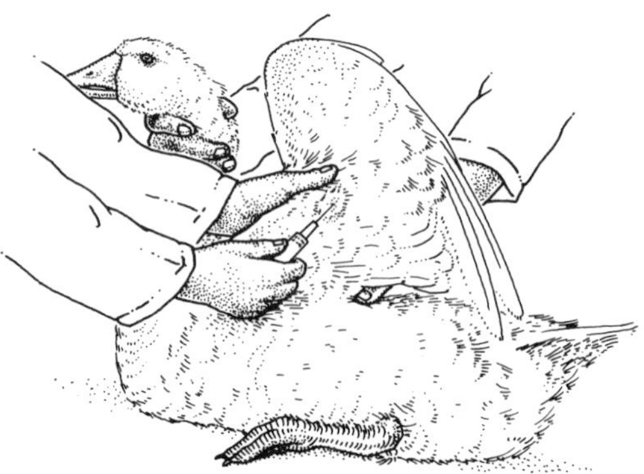

FIG. 7.22 A goose restrained for bleeding from the wing vein.

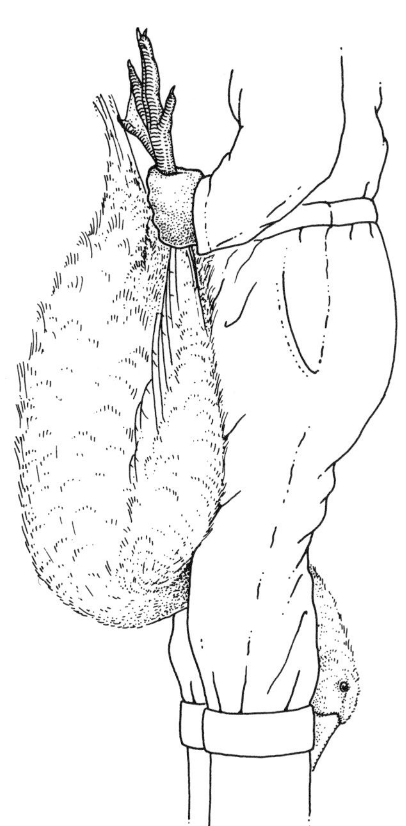

FIG. 7.23 A goose held by its feet with its head between the handler's knees.

enclosures. The goose may then be held under one arm with the wings folded and the head to the front or rear (Fig. 7.20). If the head is to the front, it may be held with the free hand to prevent injury to goose or catcher (Fig. 7.21).

If the goose is at large, it may be caught by means of a catching device like a shepherd's crook, which is hooked around the neck, thereby immobilizing the moving goose, which can then be picked up by the wings as previously described. A fisherman's keep net can also be used to catch individual geese by immobilizing them.

If too many geese are penned at one time for catching, there is a danger that crowding will occur in one corner and suffocation may result, especially if the weather is warm.

Geese may have to be caught to assess their bodily condition, for blood sampling or swabbing, to administer medication by injection or other route, for transport or for some other reason.

Blood Sampling and Swabbing

The wing vein is the vein most readily available for blood sampling and it is preferred to the jugular in this species. For blood sampling, the goose may be restrained, with both its feet on the ground, by gentle pressure on the back of the goose, and a wing raised to expose the wing (brachial) vein (Fig. 7.22).

Wing clipping

Wing clipping is employed to prevent flight. It comprises the cutting of the primary feathers on one wing only. This results in uneven aerodynamics, making balanced flight impossible.

Further Reading

Birds: Their Structure and Function, 2nd edition, 1984 (Editors King, A. S. and McLelland, J.). Baillière Tindall, London.

Sainsbury, D. (1984) *Poultry Health and Management*, 2nd edition. Granada, London.

The Management and Welfare of Farm Animals, 3rd edition, 1988. The UFAW Handbook. Baillière Tindall, London.

8

Horses and Ponies

ALISON SCHWABE

The Horse's Natural Behaviour

Anticipation is the key to the safe handling of horses and ponies. Their behaviour, though sometimes sudden and violent, can often be predicted by careful observation. Developing an awareness of the horse's natural behaviour and thus avoiding certain situations can enable handlers to prevent accidents. Most horses will behave better if handled with confidence, firmness and kindness. Only deliberate bad behaviour or naughtiness should incur reprimand and this should be immediate. Thrashing a horse or using any other form of cruelty is never justified. It will only produce fear and panic, and the horse, which is invariably stronger than the handler, will react instinctively and be less likely to perform the required activity. Wearing stout footwear is essential to give confidence to the handler (Fig. 8.1). Boots or shoes with steel toe-caps are well worth considering as they protect the foot from trampling and enable the handler to stay close to the horse without fear of injury.

The modern domestic horse retains the primitive instincts of its ancestors. It is strongly motivated to search for food and the companionship of other horses, particularly those familiar to it. This gregariousness can frequently be observed in any field (Fig. 8.2) where horses tend to rest or graze in groups. If one animal is removed, there is much frantic whinnying and even attempts to jump out and follow. Horses which appear to be enemies at home will often become devoted friends in strange surroundings, for example at a horse show. The other motivator — food — can be used to persuade an animal to do many things and to distract its attention from something it finds frightening, such as being clipped for the first time. Obviously food can only be used for a limited time but it may help with the initial calming of the animal.

The primitive Equidae were only able to survive the attacks of predators by kicking, biting, bucking

FIG. 8.1 Stout footwear for handling horses.

Fig. 8.2 Horses are gregarious by nature.

and running away. These instincts remain in the modern horse but can be modified by training. Horses rarely kick people deliberately but may do so if startled. They do not necessarily wait to find out that it is their trusted owner who has appeared suddenly and patted them on the rump. As far as they are concerned it could have been a sabre-toothed tiger! Speaking to the horse whenever you approach will give it warning and time to recognize you as a friend. The most risky situation is to get between the hind ends of two or more horses, particularly when there is competition for food (Fig. 8.3). Horses have few scruples about kicking each other in such situations and it is easy for people to get caught in the cross-fire.

Biting is also most commonly aimed at another horse but youngsters, stallions and horses that have been teased or mishandled may bite or nip. Even a playful nip can be very painful. The horse will indicate the degree of its intentions by laying

Fig. 8.3 Competition for food.

back its ears and sometimes baring its teeth. There are a few horses that will attempt to savage people by making a determined attack, even approaching from some distance away in a field. This is a rare but serious vice and such animals should be disposed of or treated with extreme caution. They can never be trusted.

The bucking instinct is mainly a problem if it is carried over into the times when the horse is being ridden. It is important to check that it is not caused by pain or ill-fitting tack. The answer is to keep the horse's head up and sit tight. Horses often appear to buck for fun, particularly when they have just been turned out into a field. It is important to keep hold close to the horse's head to avoid trouble. Sloppy handling in these circumstances is a frequent cause of accidents.

Running away is the horse's main defence and as it evolved from the small forest-dwelling *Eohippus* (*Hyracotherium*) to the plain-dwelling *Equus caballus* it became a supreme athlete. A horse's training involves familiarizing it with everyday things so that it realizes that running away is seldom necessary. Nevertheless, it is important to keep a tight hold if the horse is frightened so that it has time to assess the danger for itself. Even the most highly trained police horses can be seen to 'spook' at sudden movements and unfamiliar sights and sounds. Learning to interpret the signals a horse will give you involves observing and spending as much time as possible with them. Lucy Rees's book *The Horse's Mind* is an excellent guide to horse and pony body language.

Approach and Catching

Always approach the horse from the front and slightly to one side where it can see you. The horse's vision behind and directly in front is not good. Speak as you approach so as to give the animal warning. Handle the horse first on the lower neck or shoulder as this is a safe area (Fig. 8.4). Put the lead rope around the horse's neck. Most animals will consider themselves caught at this stage, but if not, and you are in an open space, there is little you can do to stop it charging off. Such animals should wear a headcollar in the field. The majority of horses, however, will allow you to put on the halter or headcollar. When catching a horse in a field it may help to take an edible 'bribe' in the form of an apple or carrot. If there are several horses in the field it is not advisable to take a bucket of food. The horses will all come milling about and you may get trampled or kicked. It is better to get as close a possible to the selected animal before revealing your bribe or it may be driven off by a more dominant animal. Horses reared on mints may even respond to the rustle of sweet papers. Keep the headcollar hidden until you have got hold of the horse. A hand round the nose, grasping the mane or an arm round the neck of a small pony will usually suffice.

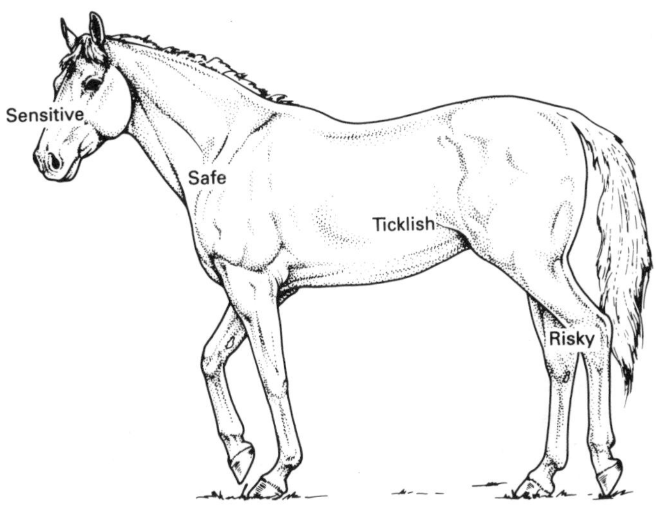

FIG. 8.4 Safe and risky handling areas.

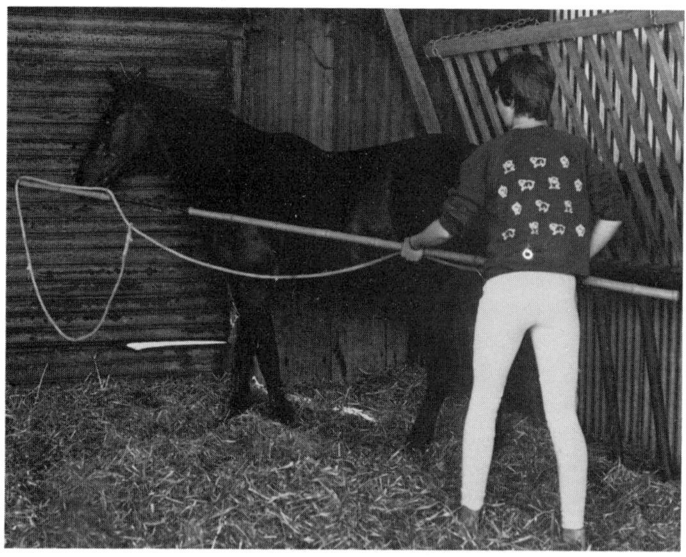

FIG. 8.5 Improvised lasso for catching a nervous animal.

There are a number of tricks for catching difficult horses or ponies which are worth trying, depending on the circumstances:

1. Offer food over the fence. Some animals will then consider you to be a visitor rather than a catcher. This works best if the animal is wearing a headcollar. It will generally submit once you have hold of the head.
2. Send someone out who does not normally ride the pony or who is dressed in non-horsy clothing.
3. For an animal which will come close, but snatches its head away when you try to catch it — crouch on the ground with some food close by. The unsuspecting animal may allow you to touch a front leg which you can then pick up. The horse then assumes it is caught. Leaving a short length of rope or leather on the headcollar may also help with a head-shy animal.
4. Cornering the horse with the help of two or more people sometimes works, providing it is not the type of animal which starts galloping off when you are still a long way away. Try to avoid exciting the animal and if possible do not let it exceed a walk. Many animals will give up once they are cornered and allow themselves to be caught. Stretching a rope between helpers may be necessary to fill the gaps.
5. Where catching is a frequent problem, it may be necessary to build a stockade in the corner of the field. Always feed the animal in there. It may be necessary to arrange a simple remote control system for closing the gate. The fencing should be high post and rail or boarded if there is a risk of attempted escape by jumping. Running an animal into a stable is another alternative.
6. For an animal that is difficult to catch even in a confined space it may be necessary to use a lasso (Fig. 8.5). The animal may be frightened, unhandled, or reluctant because it has been injured or is afraid of repeated treatments. A home-made device using a light but strong bamboo cane with two large hooks screwed into it about 1 m apart can be used in conjunction with a stout rope with an eye in one end. Make a noose with the rope and hang it off the hooks to keep it open. This can then be looped over the animal's head from a distance or over a stable door to give some degree of restraint. The cane is removed once the noose is round the animal's neck.
7. Desperate situations need desperate measures. The only alternatives are a skilled

cowboy with a lariat or a veterinary surgeon licensed to use a dart gun.

Many of the above techniques may be aided by removing other animals from the vicinity. Never chase the animal round the field. It only makes matters worse. Horses that have been well handled from birth are seldom difficult to catch, the first few hours of life are particularly important in this respect.

Headcollars and Halters

For most procedures around the stable yard, the horse can be restrained by a halter or headcollar. There are a number of different types (Fig. 8.6). Headcollars are better than halters as they cannot be slipped off easily. They may be made of leather, which is smart, strong, and expensive, or webbing which is quite adequate for everyday use and about a quarter of the price. The quality of webbing headcollars varies considerably. The stiffer type with traditional buckles and non-adjustable nosebands wear best but for growing animals it may be better to choose one with an adjustable noseband. A foal slip is a small headcollar made adjustable for this reason. It is also usual to have a small leather strap on a foal slip, to guide the foal when leading its mother.

Halters (Fig. 8.6) are cheaper still and are made from sisal or cotton rope. They can be improvised using a length of rope in an emergency but can easily be slipped off by the horse rubbing its ears. A Yorkshire halter (Fig. 8.6, centre) is a slightly more sophisticated version. It is made of coarse woven webbing and has a string throat lash which fastens in the near side to make it secure. Yorkshire halters are traditionally used for showing heavy horses and the heavier native breeds. The lead rope is usually an integral part of the halter. Headcollars should have a stout lead rope with a spring clip worthy of the rope. Horses which discover they can break a rope become a menace.

Tying Up

All horses should be trained to stand quietly when tied up. This should be achieved at an early age, as a horse which 'cannot be tied up' is a considerable nuisance. Always use a quick-release knot (Fig. 8.7). Never tie an animal to anything which might rattle, bang or fall on it. Choose a stout ring, post or rail (Fig. 8.8). There is some

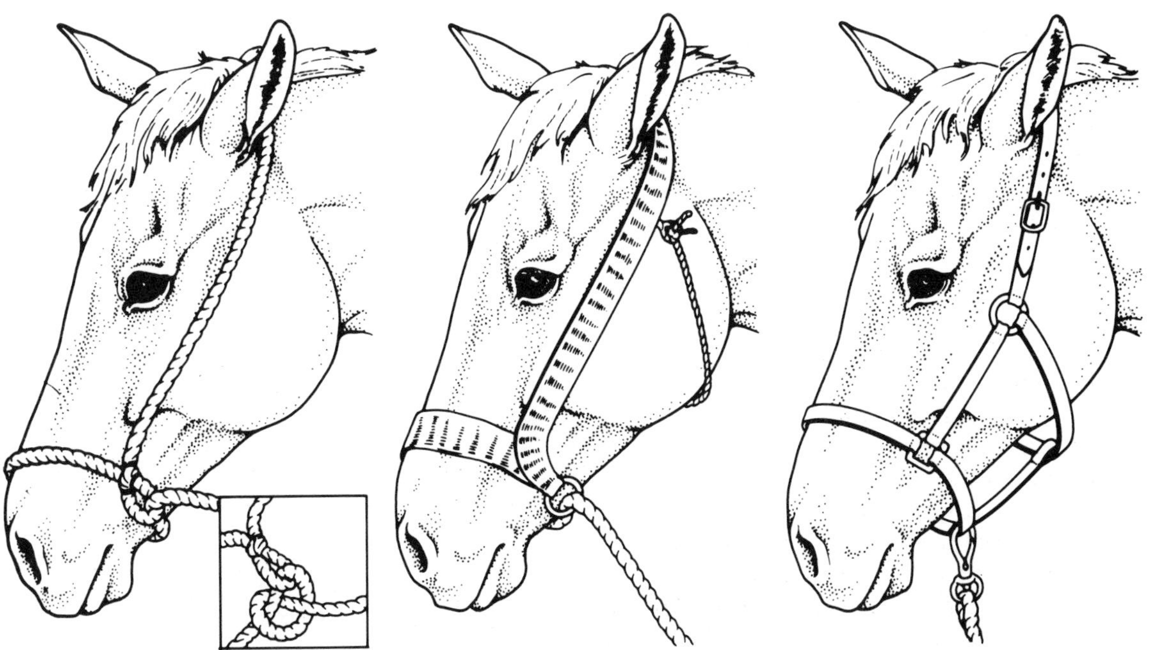

FIG. 8.6 Halters and headcollars: (*left*) rope halter; (*centre*) Yorkshire halter; (*right*) headcollar.

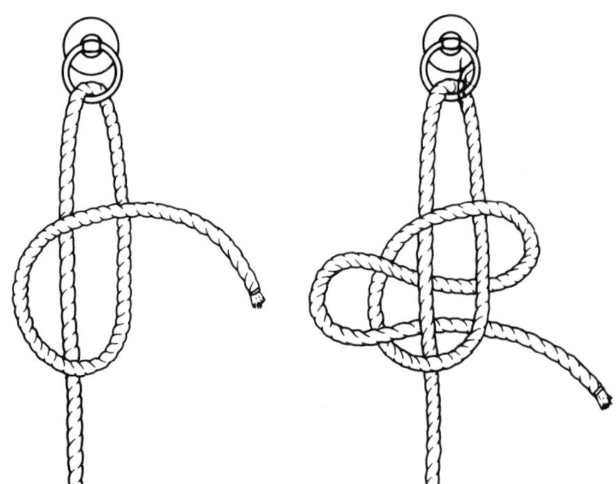

FIG. 8.7 Quick-release knot.

controversy as to whether horses should be tied to a piece of string which will break if it pulls back. There is a risk of head injury, or falling over if nothing gives way, but it can also lead to habitual halter-breaking, not to mention the dangers of a loose horse. It is probably better to train a horse with a very strong, broad, unbreakable headcollar (e.g. rawhide) padded behind the poll. The horse should not be left unattended and may be 'quick-released' in an emergency. A persistent halter-breaker should undergo re-training. The method illustrated by Sidney Galvayne in his book *The Horse*, published in 1888, works extremely well (Fig. 8.8). If the horse pulls back it pulls against itself. Never tie a horse up by the bridle. They invariably panic and break the bridle. Western horses are sometimes trained to be tied up with a bridle or ground tie, i.e. remain where they are left if the (split) reins are left hanging down. This training requires considerable patience.

Leading

Horses are traditonally led from the near or left-hand side (Fig. 8.9). They should, however, be trained to expect approaches from both sides. When being led on the road, the handler should be on the off side, between the horse and the traffic. The lead rope should be held near the horse's head with the spare rope held in the other hand. The handler should walk beside the horse's shoulder. The rope should never be wound around the hand in case the horse takes off. The handler could be unable to let go and be dragged. If this is a possibility, the horse should wear a bridle. All lively or difficult horses should wear a bridle for handling and all horses being led on, or near, roads should wear a bridle. It could be considered negligent not to use one. This may be the horse's normal bridle, with the reins taken over the head, or an in-hand bridle (Fig. 8.10) with a straight-barred bit which gives more control. A Chiffney anti-rearing bit (Fig. 8.11) is a very useful device which can be slipped on over a headcollar to give better control. Animals which are too young to be bitted can wear a lungeing cavesson with lungeing rein which gives better control than a headcollar but the youngest can be introduced to a nylon foal bit at an early age (Fig. 8.12) if being shown in-hand.

FIG. 8.8 Re-training a halter-breaker — Galvayne's method.

FIG. 8.9 Leading in hand.

A lungeing rein or a longer-than-normal lead rein is useful for horses which are likely to 'play-up'. (Fig. 8.13). If, for example, the horse should rear, the handler can keep clear of the flailing hooves but still keep hold of the animal. It is important to get back close to the horse's head as soon as possible, otherwise the handler could be kicked. Additional control can be provided by a roller and side reins. This is commonly used for stallions which are particularly inclined to bite and rear — albeit in playful exhibitionism.

If a horse is reluctant to move forward, do not tug at it or stare it in the face. Stay beside the horse, give the command 'walk on' and, if necessary, tap it on the side with a stick held in the hand

FIG. 8.10 In-hand bridle.

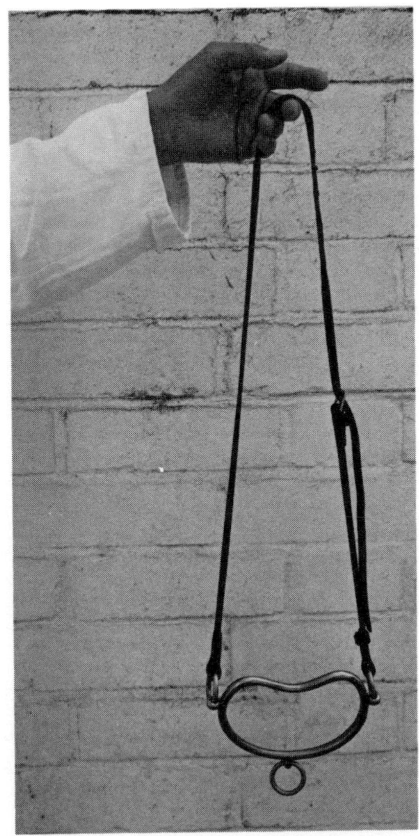

FIG. 8.11 The Chiffney anti-rearing bit.

furthest away from the horse. Alternatively enlist the help of someone who can go behind the horse with a long whip or a broom, which will enable them to keep out of kicking range. Using commands such as 'walk on', 'halt' and 'stand' from an early age can be very helpful.

Where a horse has become a problem to lead forward and its inclined to back off or throw its head in the air, an improvised rope halter may be constructed to discourage this. This halter was developed by Monty Roberts, the well known American trainer of horses, and is referred to by him as the 'come-along line'. A piece of rope about 3 m long and about the thickness of a washing line should be used to construct the halter (Fig. 8.14). If the horse pulls back, the rope presses behind the poll and tightens around the nose. The horse quickly learns to avoid this by moving forward. This improvised halter is also useful if you come across a loose horse and can only find a piece of rope or baler twine to restrain it. It is also very useful for loading difficult horses.

Fig. 8.13 A long rein is required for a rearing horse.

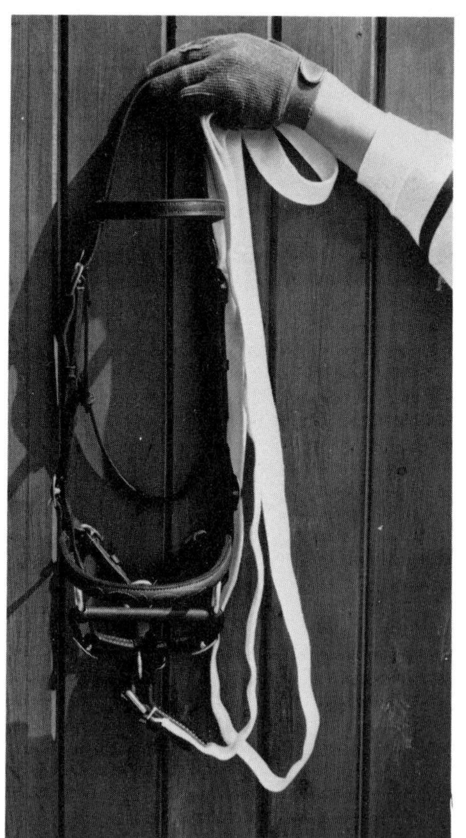

Fig. 8.12 In-hand bridle with nylon foal bit.

To make a horse stand still, particularly one which is fidgeting, stand facing its head, slightly to one side to avoid possible striking out from a front leg, and hold each rein close to the bit on either side of the head or hold both sides of the headcollar. The horse may then be kept straight and be prevented from moving forward by a gentle tug on the bit. Holding up a front leg will also help, for example, if someone is trying to bandage a hind leg (Fig. 8.15).

Picking Up Feet

Most horses have their feet picked up and cleaned out daily and so should not present a problem. Speaking to the horse, approach at the shoulder and, facing the tail, run the hand nearest the horse down over the elbow, back of the knee and tendons to the fetlock. Tug at the fetlock and give the command 'up'. As soon as the horse lifts its foot, catch the hoof by the toe and support it with the other hand. It is important to ensure that the horse is standing reasonably square to begin with and that the limb is well supported once lifted. The horse will then be unlikely to resist. Do not allow a horse to get in the habit of snatching its limb away and stamping it down. Your toes may be in the way and your fingers could be cut by sharp clenches in the hoof. To lift a hind limb (Fig. 8.16) run the same hand along the horse's back, the rump and down the back leg, to the point of the hock. Move your hand round via the anterior aspect of the cannon bone to the medial aspect of the fetlock joint. Tug at this with the command 'up' and encircle the hoof with your hand and support it. Keep the limb well under the horse but slightly out to the side so as to avoid unbalancing it. If the horse resists lifting a foot, lean into it to push its weight on to the opposite limb and in the case of a forelimb, your elbow into the back of its knee (carpus). Horses which have had their feet regularly picked out in a particular order will generally oblige by having the next one ready for you.

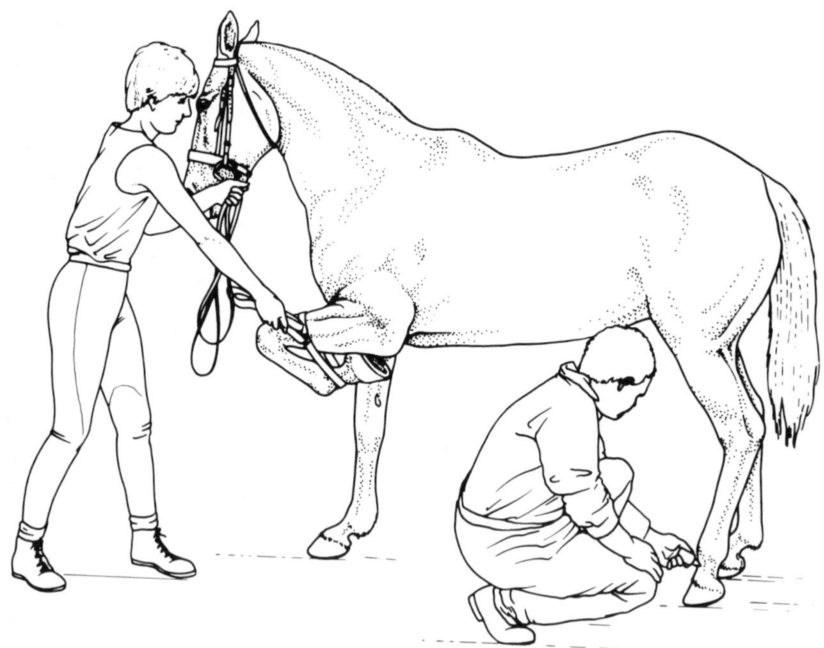

Fig. 8.15 Holding up a front leg to prevent kicking.

Manipulation for Showing or Investigating Lameness

'Trotting up' is the name given to showing off a horse's paces when showing in-hand and also to showing the horse's paces during an investigation for lameness. It is important that adequate space is available for this. For lameness investigations, both hard and soft surfaces are useful. The horse is first led away from the observer in a straight line, for a distance of about 25 m, turned and walked back and on past the observer. The procedure is then repeated at the trot. Except when turning, the lead rope should be allowed to lie over an open hand (Fig. 8.17) to avoid interfering with the head movement but in readiness to grasp a firm hold should the horse play up. In normal circumstances the horse should be turned away from the handler to lessen the likelihood of trampled toes and to keep from obstructing the view of the horse. However, in some cases the veterinary surgeon may wish to see the horse turned in both directions. In confined spaces it may be necessary to turn the horse towards you. Remember to keep your feet out of the way. A veterinary surgeon may also require a horse to be turned sharply to assess its willingness to cross its feet, or put weight on them. This is achieved by facing the middle of the animal, pulling its head towards you with one hand and chasing its quarters away fom you with the end of the rope or reins in the other hand (Fig. 8.18). This is generally done in both directions. It may also be necessary to back the horse. Tug on the rope or reins with one hand and push a fist into the breast giving the command 'back' (Fig. 8.19). Be prepared for a trained carriage horse to move back smartly just on command.

Lungeing

Lungeing (Fig. 8.20) is a useful way of exercising a horse which cannot be ridden, as well as being a useful part of training. It can be used to assess the effects of exercise on lameness or on the circulatory or respiratory systems. It is a very intensive form of exercise and should not be used on yearlings and only with caution on two-year-olds to avoid undue strain on limbs. It is recommended that the handler wears gloves and a hard hat. The horse should be protected with brushing boots. It is helpful if the horse has been trained to words of command whilst being led in hand. When lungeing to the left the lungeing rein should be held in the

FIG. 8.16 Lifting a hind leg.

FIG. 8.17 Trotting up for lameness investigation.

left hand and the whip in the right hand. The slack of the rein may be held in either hand but in large loops for easy adjustment, and not wound around the hand. The handler should never be in front of the horse's shoulder and should walk in a small circle. Pointing the whip at the horse's elbow will

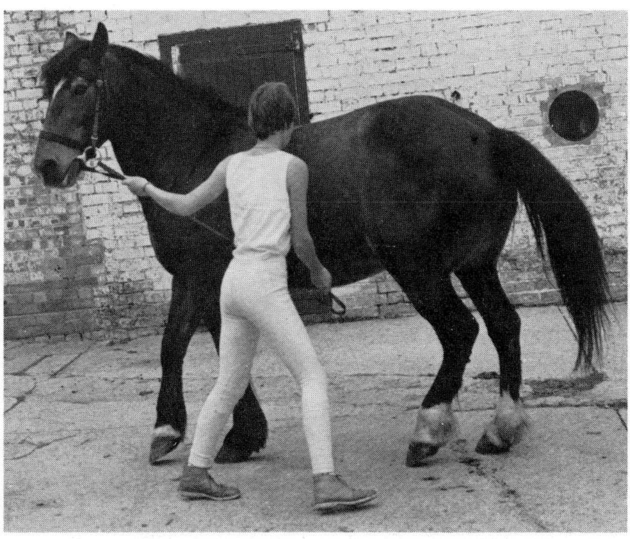

FIG. 8.18 Turning a horse sharply.

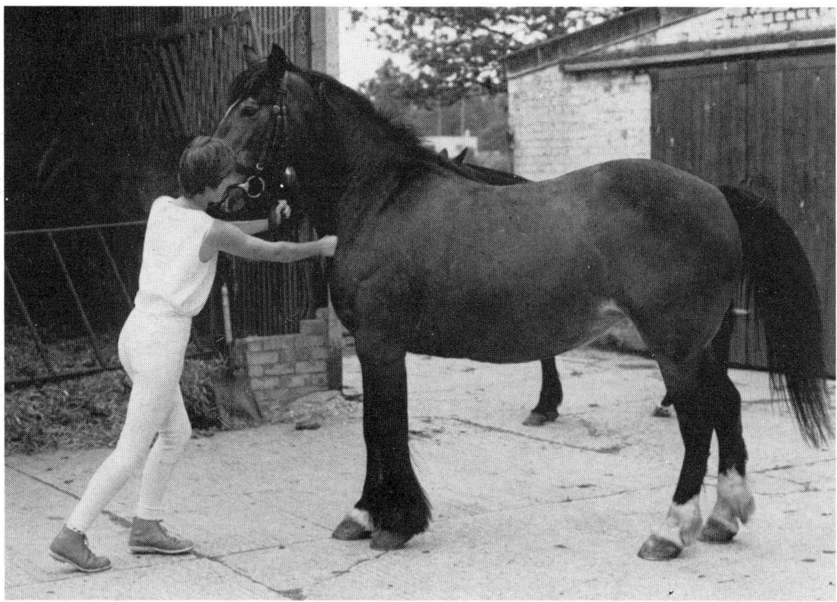

Fig. 8.19 Backing a horse.

keep it out on the large circle. Tugging gently at the lunge rein will slow the horse and a flick of the whip behind it should send it on if it does not obey words of command at once. Pace and direction should be changed frequently. Always halt the horse on the circle and do not allow him to turn in towards you. If this is a problem, a second rein passed through a surcingle ring on the outside and round the quarters will help to keep him straight. It is not a good idea for an inexperienced horse and handler to learn this technique together!

Handling Groups of Horses

The gregarious nature of horses should be born in mind when handling groups of horses, but

Fig. 8.20 Lungeing.

strange horses should not be allowed too close to others. When leading or riding in a string, keep at least one horse's length between animals to avoid kicking. Horses being introduced nose-to-nose for the first time will often squeal, and strike out with a front foot or even try to turn quickly and kick. These possibilities should all be anticipated by the handler, consideration for the safety of others is all-important and those with most control should keep an eye on any handler who may be having difficulty. When crossing a road it is vital that all horses cross together. One person should take charge and check that all are ready to cross simultaneously or, in larger groups, stand in the middle of the road to hold up the traffic and see the whole string across. If one animal is left behind it could become out of control, try to charge across and cause an accident.

When turning a group of horses out into a field, similar precautions should be taken. All the horses should be taken into the field and held well apart and clear of the gateway. The gate should then be closed. They should all be turned to face the gate and, when everyone is ready, be released together. This is because horses being turned out together will often be keen to be released and will buck and kick and gallop off. One person could be left struggling with a headcollar on an over-excited horse. Turning the horses to face the gate means the horses have to turn before they charge off and the handlers have time to step clear.

In any situation where one person, riding or leading, loses control, the others should try to remain calm and stand still. Trying to rescue a bolting horse heroically at a gallop will only make matters worse.

Prevention and Avoidance of Kicking

It has already been mentioned that horses seldom kick people deliberately but you should never put yourself within kicking range of an unknown horse. You should also be aware of the position of other people, particularly children, who might wander behind animals. It is, however, permissible, to go behind a known horse, if it is away from others, provided that you make sure that it knows where you are beforehand. Keep a hand on its hindquarters and walk very close to it so that you can anticipate its movements. Holding the horse's tail and leaning on it slightly will discourage it from kicking. Remember that horses can also cow-kick (Fig. 8.21) forwards, particularly if irritated by a fly or by the girth being tightened. Some may strike out forwards with a forelimb (Fig. 8.22) which can be dangerous and should be firmly discouraged.

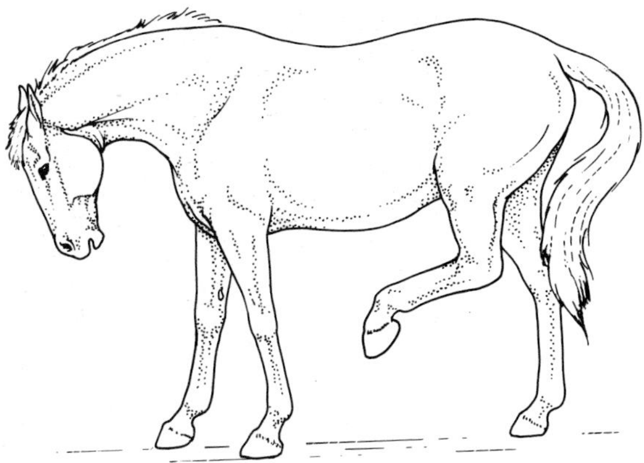

Fig. 8.21 Cow-kicking.

If it is necessary to go within kicking range of a horse which is not trusted completely, for example, to put on a tail bandage or carry out a veterinary procedure or examination, precautions should be taken. The simplest method is to hold up a front leg; this can frequently be done by the person holding the horse's head. The leg on the same side as the procedure is being carried out should be lifted as it is more difficult (but not

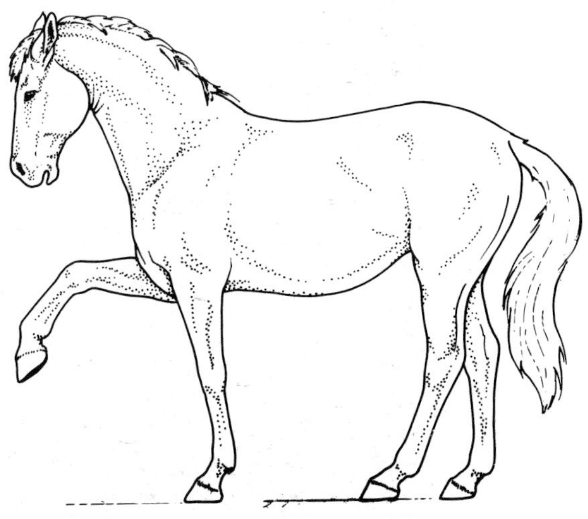

Fig. 8.22 Striking with a foreleg.

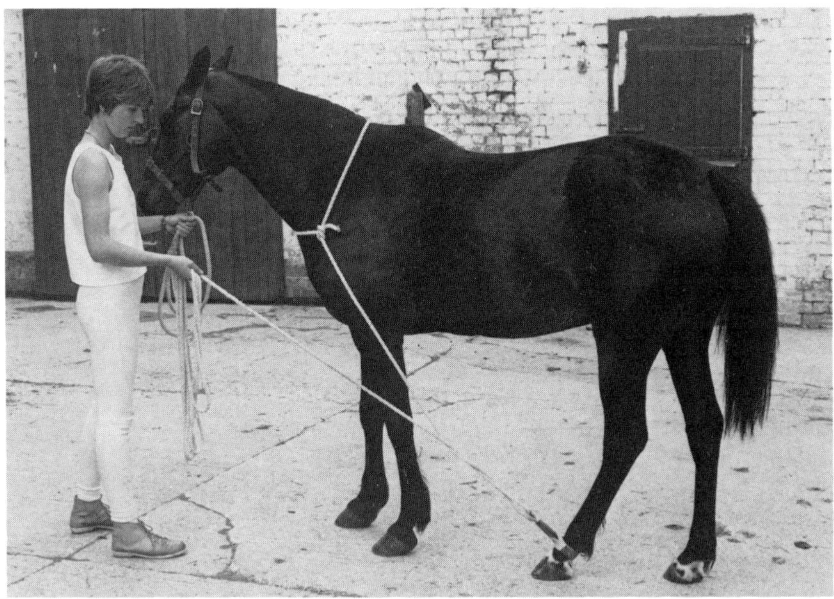

FIG. 8.23 Side-line to prevent kicking.

impossible) for a horse to stand on only two legs on the same side at once. If the leg is to be held for any length of time, a leather strap such as a stirrup leather, can be used to hold the leg up. This puts less strain on the handler and causes no discomfort to the horse (Fig. 8.15). The leg can be released instantly if the horse tries to go down. A protective knee boot should be worn by the animal if this is a risk on a hard surface. This is a useful technique when a mare is being covered. Once the stallion has mounted her, the leg can be released to allow the mare to stand on all four feet. The mare handler can also keep well clear of the stallion's hooves.

A side-line (Fig. 8.23) is a quick and easy way of preventing kicking. It is much used in some countries and by the military. A rope is tied around the horse's neck using a bowline (Fig. 8.24) to prevent the knot tightening on the horse. The long end is passed through a hobble on the hind pastern or alternatively round the pastern itself, provided it is protected by a bandage. The end of the rope is then returned to be held by the handler after taking a twist round itself. If the horse lifts the leg to kick, it is unable to get any backwards force. Service hobbles are, in effect, fixed, double side-lines which prevent a mare kicking a stallion but, unless a horse is used to them, it may panic at such restraint and fall over.

The alternative to stopping a horse kicking is to use a barrier for protection. This is probably the best solution in cases where it is necessary to spend some time behind the horse, for example, when carrying out gynaecological examinations.

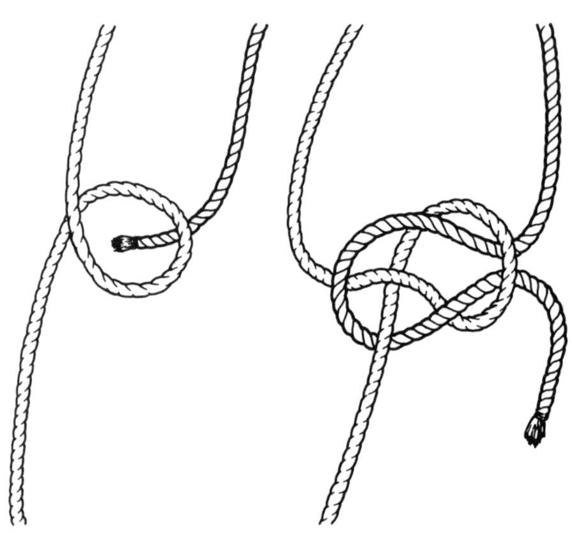

FIG. 8.24 Bowline knot.

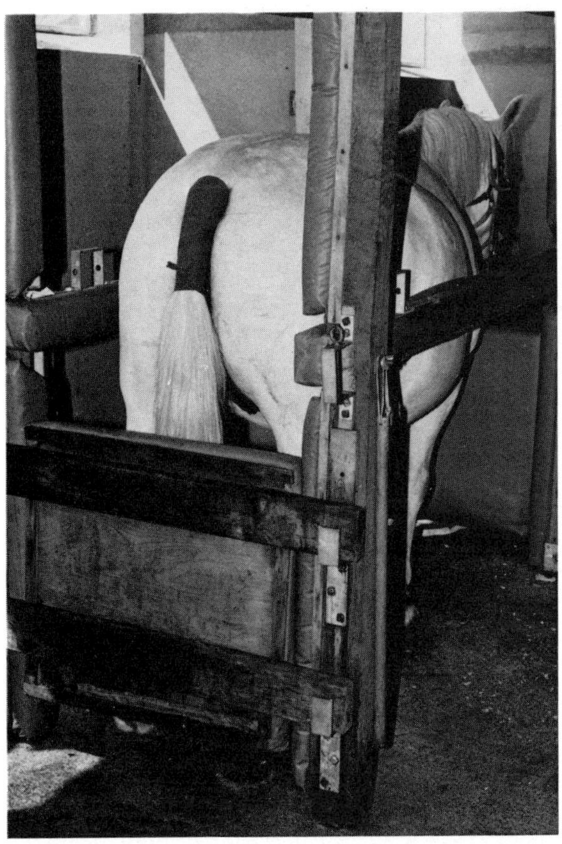

FIG. 8.25 Horse stocks.

Purpose-built horse stocks (Fig. 8.25) should be installed where this is done frequently. They should be made of hard wood, have padded sides and padded removable boards at both ends. If stocks are not available, they may be improvised by the use of bales (Fig. 8.26). Another option is to stand the horse in an open stable doorway to limit the 'target area'. The veterinarian can then retreat behind the wall if the horse gets agitated. He is still vulnerable when carrying out the main examination. It is very dangerous to carry out rectal examinations over a stable door or gateway. If the horse sinks down, as occasionally happens, an arm could be broken, and the horse could receive internal injury.

Additional Restraints

Additional means of restraint are sometimes needed for frightened or injured animals in order to carry out examinations or treatments. For this purpose a twitch may be used. There are two types. A traditional twitch (Fig. 8.27) is made of a short pole 50–70 cm in length with a loop of rope 6–7 mm thick at the end. The loop, which should be 40–50 cm long, is twisted round the horse's upper lip. The best technique of applying the twitch is to put the thumb and fingers — *except* for the index finger — through the loop and grasp the upper lip (Fig. 8.28). Slide the loop off the hand on to the lip and twist it up firmly. This method will prevent your fingers becoming caught in the loop. Originally it was thought that the twitch caused pain which distracted the horse's attention and there is no doubt that this is partially the case. After a few minutes, however, the horse often appears sedated. Work at the University of Utrecht has shown that pressure on the upper lip causes the release of natural endorphins which may have an anaesthetic effect similar to that induced by acupuncture. Nevertheless, the twitch should be kept on only for the minimum time necessary and never for longer than 5 minutes, less if it is applied tightly. The 'humane' twitch (Fig. 8.27) is a commercially available device with a long-type action. It is probably slightly easier to apply and less likely to damage the skin. Twitches should never be applied to ears as permanent damage can easily be caused. However, the head may be steadied by placing one hand half-way up the nose and grasping the ear round its *base* firmly (Fig. 8.29). A method of distracting the horse's attention briefly whilst, for example, giving an injection is the 'gypsy hold' (Fig. 8.30). Grasp a thick fold of the skin of the neck and give a slight twist sideways.

Examination of the Mouth

It is fortunate that horses have a convenient gap (the diastema) between the corner incisors and the first premolars. This enables the mouth to be easily opened without the handler being bitten. If all that is required is a quick look at the labial surfaces of the gums or teeth, most horses will tolerate the lips being drawn back. In exceptional cases, a twitch may be necessary. Many horses will also allow a good view of the mouth without resorting to a gag. One hand should steady the nose while the other is inserted through the diastema, palm downwards, to grasp the tongue. The tongue may then be

Fig. 8.26 Improvised stocks.

drawn out sideways, which keeps it out of the way, or rolled upwards and backwards so that the fist forms a 'gag' and keeps the mouth open. The tongue can be slippery and difficult to hold. It should *never* be used to pull the horse's head down or sideways. An oral dose of medicine may be administered in this way.

With procedures such as tooth-rasping, it is probably easier to use a gag. This will be secure and, once inserted, leaves the hands free to carry out other procedures. The two most commonly used types are Swale's gag (Fig. 8.31) and Hausmann's (Fig. 8.32). Swale's gag acts by wedging the molars apart. The headcollar or halter noseband must be sufficiently loose. It is best to have the horse backed into a corner to prevent it running

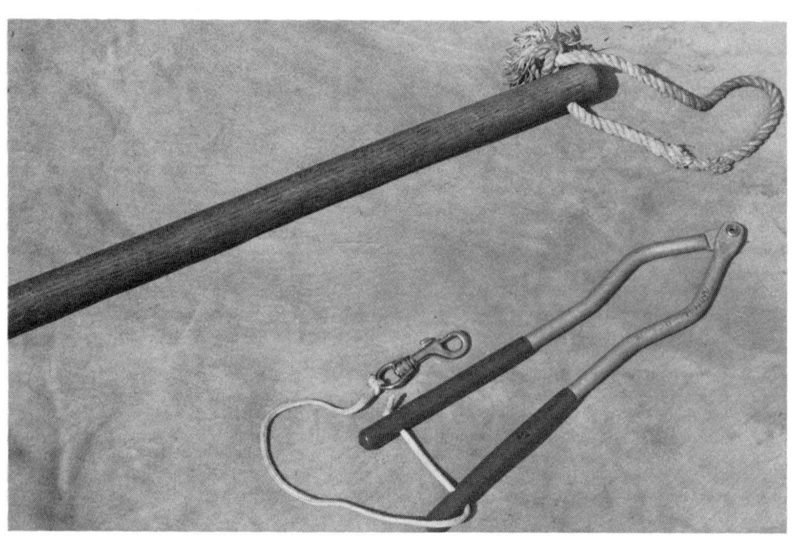

Fig. 8.27 The traditional twitch (*top*) and the humane twitch.

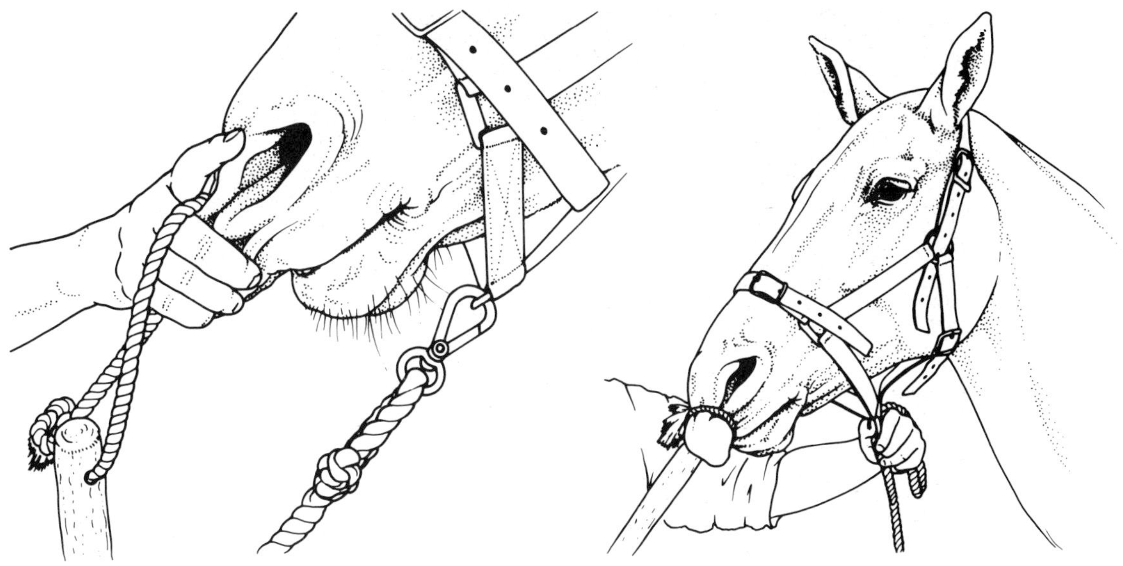

FIG. 8.28 Application of the traditional twitch.

back. Stand in front of the horse, steadying the head with one hand, and holding the handle of the gag with the other. Gently push the spiral through the diastema and upwards and backwards until the horse is biting on the metal spiral with its molars. With your thumb in the leather strap, grasp the headcollar with the remaining fingers to ensure the gag cannot fall out. Alternatively buckle the strap to the headcollar. The gag may be inserted on either side of the mouth. To remove the gag, release the leather strap and allow the horse to spit it out.

Hausmann's gag is heavier, expensive and more cumbersome to insert but once in position it is very

FIG. 8.29 Steadying the head.

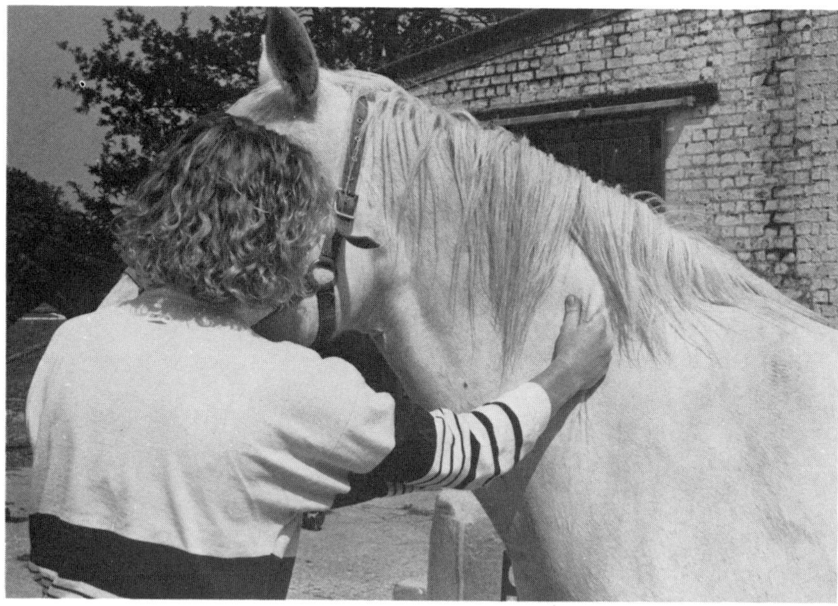

FIG. 8.30 The gipsy hold.

secure and leaves the molars free. It is helpful, but not essential, to have two people to put it in as it is inclined to fall apart easily, due to poor design. Recent variations on Hausmann's gag have improved on the design. The gag is put on rather like a bridle but with the two incisor plates positioned so it may be opened by pulling apart at the incisor plates, one ratchet at a time. To close the gag, squeeze the thumb presses simultaneously. Do not leave the gag in longer than necessary. Most horses will tolerate gags without much resistance. Hausmann's gag comes with rubber covered

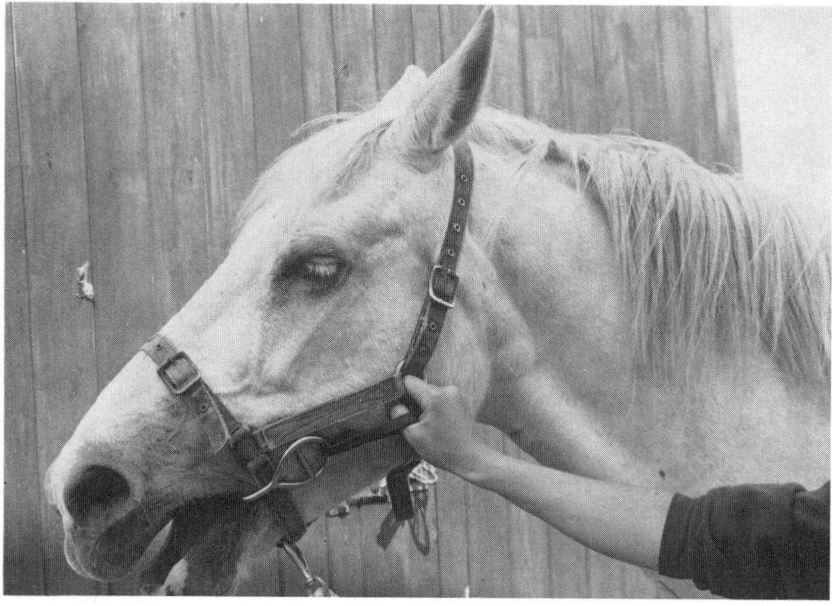

FIG. 8.31 Swale's gag in position.

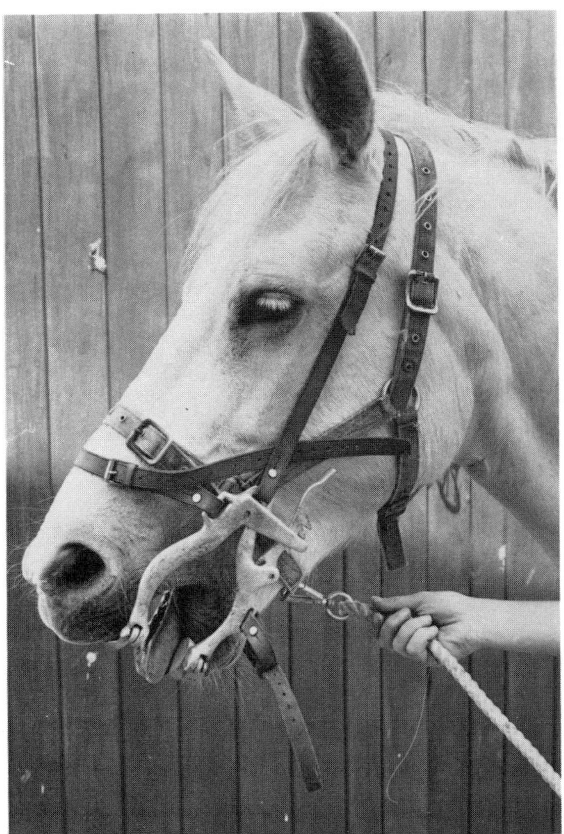

FIG. 8.32 Hausmann's gag in position.

bars as an alternative to the incisor plates. These are primarily intended for use on cattle but may be used in the diastema of the horse if access to the incisors is required. Varnell's and Hitching's gags (Fig. 8.33) are rarely used nowadays but serve a useful purpose when the horse is anaesthetized in allowing intubation. They are rather cumbersome in the conscious horse.

Handling Foals and Young Animals

Foals and young animals tend to learn a good deal by their mistakes. It is important that the lesson should not be too hard and the animal injured. The young foal's instinct is to follow its mother closely and, as it gets older, it will range further and further on its own. The foal should be handled frequently from birth and wear a foal slip so that it becomes used to being led and handled about the head. In many studs, mares and foals are turned out during the day and brought in at night. At first two people are necessary — one to lead the mare and the other to push the foal along from behind and to guide its head with the other hand. Very soon it will be possible for one person to lead both (Fig. 8.34). It is important to ensure that the mare and foal do not become separated, and so special care should be taken at gates and doorways. If the foal finds itself on the opposite side, it may panic and try to jump or force its way through. Most studs have extra-wide gateways and stud fencing and sometimes rollers on the sides of doors. The foal should be sent through first. If very young foals are unable to stand they may have to be carried. This is done by grasping the foal round the brisket and below the rump (Fig. 8.35). Considerable strength may be needed.

Any animal unable to stand should have thick bedding and be turned frequently to prevent pressure sores. If the animal is likey to be recumbent for some time, consideration will have to be given to lifting with slings (Fig. 8.36). Not all patients settle to these and all need constant nursing to prevent pressure sores developing. Nevertheless, many animals with serious fractures have fully recovered after several weeks on slings. The slings should be adjusted so that the horse's feet all touch the ground and it can take part or all of its own weight if it chooses. Bales stacked in front and behind may help to prevent it attempting to move or struggle. The animal should be allowed to lie down for part of each day.

Horses which are to be anaesthetized need an experienced handler who will not panic if an individual shows a bad reaction to the anaesthetic and will help to ensure it goes down gently in a satisfactory position. The working area should be a padded recovery box, an open field or a large area with deep litter, straw or woodshavings, to prevent injury. The horse should be prevented from pitching forward and encouraged to sink backwards whilst keeping a tight hold of the head to prevent it falling heavily. Once the horse is down, its limbs should be hobbled together and the handler should be prepared to kneel on the head until the horse is safely intubated and on closed circuit anaesthesia. A lightly-anaesthetized animal falling and thrashing about is very dangerous. Kneeling on the head to prevent a horse rising may be useful at other times, for example if a horse has suffered a severe

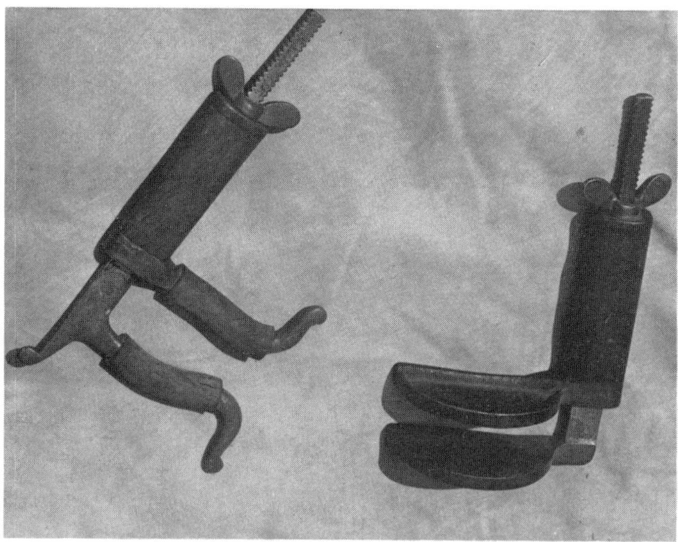

FIG. 8.33 (*left*) Varnell's gag; (*right*) Hitching's gag.

injury or become trapped in an accident. It is vital to keep the person's weight on the head and not just the neck. Horses are able to throw off quite heavy weights from their neck if any movement is allowed. At one time horses were frequently cast

FIG. 8.34 Leading a mare and foal.

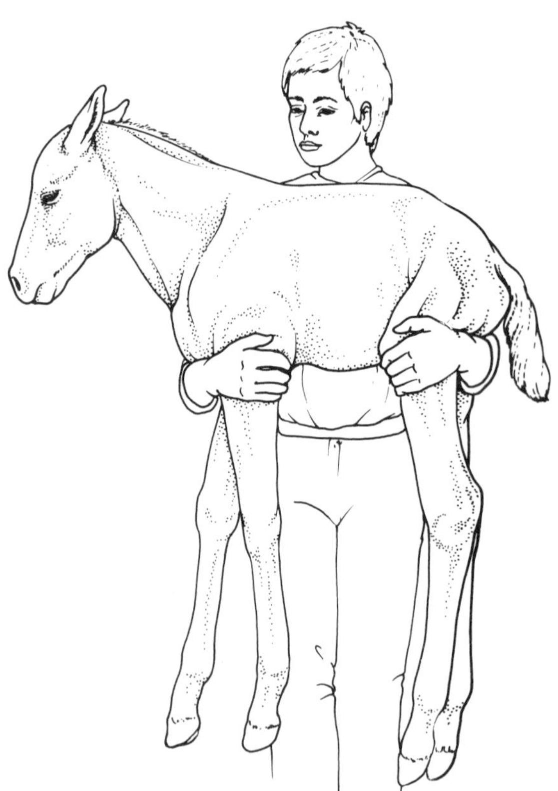

FIG. 8.35 Carrying a very young foal.

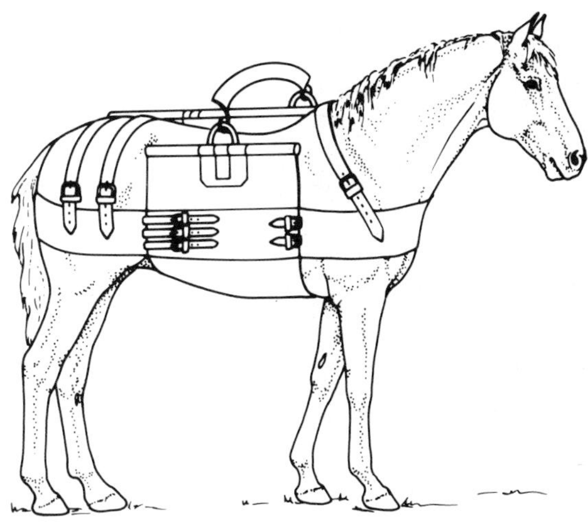

FIG. 8.36 Slings.

for operations such as castration using ropes and hobbles. With the advances made in chemical anaesthesia, the risk of injury when casting no longer justifies their use.

Transport

Many frustrating hours are spent and wasted all over the country by people struggling with difficult loaders. Much of this can be saved by early training, careful preparation, and good driving when horses are travelling. It is best to start getting the animal used to a truck or trailer as young as possible and before a journey is planned. If a mare is an easy loader, the foal can be allowed to follow her on and off the vehicle and young animals may be fed in and around it. They should not be forced on but he allowed to investigate and judge the safety for themselves. They should not be shut in to begin with. Later they may be shut in for a short time and fed. When their confidence has been built up, bang on the sides of the trailer and rock it around gently at first, so that the animal becomes accustomed to unusual sounds. It may then be taken on a short journey. Time taken with this training is well spent.

Never allow the horse to come off the trailer or truck as soon as the ramp is down. Make it a habit to insist that the horse stands for a few minutes. This will prevent it learning to rush backwards to get off. This is a frequent cause of injury to horses and people, and can damage equipment. If a horse has already developed this bad habit, the use of a front-unload trailer may help. Although breeching-straps are meant to prevent horses backing out, the rarity of functional ones on trailers is probably an indication of their lack of strength and effectiveness.

Many people will not have the opportunity to train their own youngsters and will find themselves with an adult animal which has already been frightened and learned to take evasive action when being loaded. These animals should ideally be taken back to the beginning but there are a number of useful techniques for loading difficult horses. Reliable help is essential and although food and kindness are the first line of action, there is no point in spending hours feeding nuts to smug-faced ponies standing firmly on the ramp and taking the occasional step forwards and the occasional step back.

The best method is to park the truck or trailer in an alleyway or up against a wall, so that there is no escape round one or both sides. One person should hold the head with the animal wearing a bridle or Chiffney, preferably with a lunge rein attached. Place a Newmarket loader, purchased or improvised, on the horse's quarters and have two reliable helpers, one holding each handle of the loader (Fig. 8.37). The two helpers should pull forwards as evenly as possible and not give up. The person at the head should guide the animal in, but should not pull too hard or the horse may be inclined to throw its head up. The pressure on the loins as well as on the buttocks appears to prevent the horse from both rearing and kicking, although it may produce a small buck on the first occasion it is used. Almost all horses will load by this method and after a few times the loader will not be necessary. The horse should be rewarded as soon as it is in the box. Although some of these precautions may seem unnecessary, it is much better to take them all at the beginning so that the horse learns immediately that there is no alternative to going into the box. The improvised halter illustrated in Fig. 8.14 may also be useful.

There are a number of other helpful ruses to get horses to load. The decision on whether to adopt a no-nonsense approach is a matter for judgement based on knowledge of the animal's temperament.

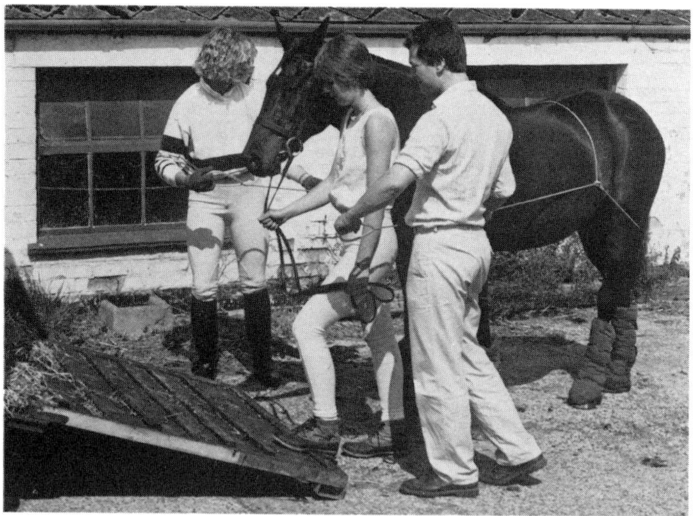

FIG. 8.37 Method of boxing using a Newmarket loader.

1. Move partitions to make the compartment wider and look more attractive.
2. Open doors or ramps at the far end and practise walking the horse right through if possible.
3. Place the horse's front feet on the ramp to get the horse started. Do not waste too much time on this technique.
4. Go behind the horse with a lungeing or hunting whip or a yard broom. Keep out of kicking range.
5. Take a strong lungeing rein or rope from the headcollar, through a ring at the front of the box, round behind the quarters and pull.
6. Have two people link arms or pull with a thick rope behind the horse.
7. Put a tight standing martingale on the horse to discourage it from rearing or throwing up its head.
8. Blindfold the horse. Some will become disorientated and go straight in, others will refuse to budge.
9. Load a companion first. This is not always convenient and in some cases puts the horse off.
10. Apply a twitch. This may help with a horse which is behaving dangerously but should not be used as routine.

When transporting mares and their young foals together, they should be allowed to see and have contact with each other but there should be no risk of the foal being squashed or trampled. A good method is to partition off an area in front of the mare. A large front and rear-unload trailer is ideal. The front space may be partitioned off with plywood and the foal may be loaded and unloaded via the front ramp and the mare via the rear ramp.

When unloading animals it is best to have at least two people present. One should untie the head and, when ready, ask for the ramp to be lowered. As previously mentioned, the horse should never be allowed to rush out either forwards or backwards. With a well-trained horse, one person can manage.

In order that this careful training is not wasted, great care should be taken by the vehicle driver to ensure that the horses have a comfortable journey. The most important factors are slow and careful cornering and gradual acceleration and deceleration. Anticipating possible traffic situations should avoid the necessity of sudden braking; engaging a low gear on hills will reduce the risk of snaking or jack-knifing. The horse should be properly protected with travelling boots or bandages. Treads — self inflicted or from another horse — are the main risk. The poll should be guarded if there is any danger of the horse hitting its head. The tail is usually bandaged for cosmetic reasons to protect it from rubbing. The horse should be tied up to a piece of string which will break if the horse goes down. On a long journey, stop at least every two hours to offer water and, where the horse is not to be exercised immediately on arrival, give access to a haynet. This should be tied well up out of the way of feet. Horses which have regular comfortable journeys will come to regard the horsebox as 'home' when away at events. Horses which have had a stressful journey have been known to lose as much as 20 kg of their body weight.

The most important factors in the successful handling and training of horses are confidence, firmness and consistency. It is the responsibility of owners to ensure that their animals receive the best possible training. It is the veterinarians responsibility to ensure that they can give clear instructions to anyone asked to help with manipulation or restraint. By anticipating problems, the safety of all concerned may be protected.

References

Galvayne, S. (1888) *The Horse*. Thomas Murray, Glasgow.
Rees, L. (1984) *The Horse's Mind*. Stanley Paul, London.

Further Reading

British Horse Society and the Pony Club (1988) *The Manual of Horsemanship*, 9th Edition. Threshold Books, London.
Edwards, E. H. (1963) *The Saddlery; Modern Equipment for Horse and Stable*. Allen, London.
Hickman, J. (ed.) (1988) *Horse Management*, 2nd Edition. Academic Press, London.
Rossdale, P. (1976) *Seeing Equine Practice*. Heinemann, London.
Schwabe, A. E. (1986) *Manipulation and Restraint of the Horse*. (Videotape). Cambridge.

9

Donkeys and Mules

JOHN FOWLER

Introduction

The domestic European donkey is generally a placid creature, whose docility makes for easy handling. Its nature and basic instincts are, however, often different from that of the horse or pony and an understanding of these differences, when applied to practical handling, facilitates the easiest partnership between man and animal.

The earliest archaeological evidence of donkeys has been found in Nubia, the North West corner of the Sudan. It is to this type of arid, semi-desert country that the donkey is physiologically adapted. As donkeys spread, over hundreds of years, around the Mediterranean coast to Europe, little has been lost of their specialist adaptation to desert-living. Similarly, the social behaviour of donkeys, necessary for survival in semi-desert conditions, remains the prevailing instinct in domesticity.

The donkey 'society' in the wild is based on a small family unit. This usually consists of a mare accompanied by her foal and yearling. Some stallions claim sexual territories over which they have exclusive mating rights and announce their ownership with very loud brays. Other stallions and colts form bachelor groups. Males in these bachelor groups only mate with mares that are not on a stallion's territory. In each of these family or bachelor groups, pair- or triple-bonds become a strong feature. These bonds between individuals are also found when donkeys are kept in groups in domestic conditions.

In their natural environment the donkeys have become, of necessity, well adapted to a very sparse diet. They function well on a diet which is high fibre, high dry matter and low nutrient concentration. Part of this adaption is an extremely economical energy expenditure at the slower paces (walk and slow trot). All these features (i.e. pair-bonds, dry conditions, high fibre diet, preference for the slower paces) should be respected and fostered in management of the donkey.

Moving

Because of their strong sense of community, any attempt to separate and remove one donkey from a small group kept together in one place will often meet with resistance from that donkey. The 'shepherd's approach' is usually more successful. It consists of moving the whole group to the new location, separating off the required donkey and then moving the remainder of the group back to the old location. In this situation it is often easier to 'herd' the donkeys rather than lead them.

If one donkey is to be separated from a group for any length of time, it is wise to sort out his friend and detain him also, thus respecting their pair-bonding instincts. (This applies equally to mares, geldings or stallions.) If, for example, an ailing donkey were stabled for treatment without his 'pair', the successful outcome of the treatment could be prejudiced by the mental anguish of separation. Similarly the otherwise healthy partner could develop evidence of stress, such as hyperlipaemia, induced by the upset of the separation. It is better to bring in *both* donkeys and, by preserving the pair-bond, protect the well-being of both.

The pair-bond is also important when a donkey dies. The surviving partner must be allowed to spend a minimum of ten minutes communing with its dead friend. Little or no pining will then result.

Catching

Donkeys are very inquisitive and appreciate company. If kept in small groups (two to five) and regularly handled from birth, they will usually approach any person who enters their field or

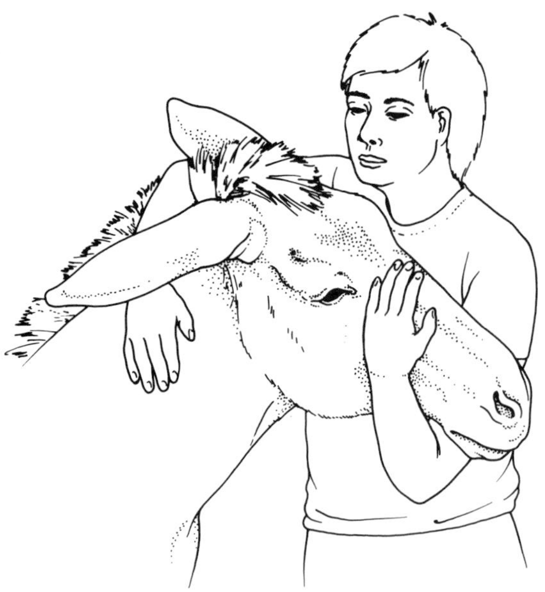

FIG. 9.1 Simple restraint of a donkey by a 'head lock'.

compound. Unless they have previously been frightened or abused, they will nuzzle and rub against the visitor and are thus easily caught.

They can be held in a simple, comfortable head lock — one arm under the chin of the donkey, the other over the neck (Fig. 9.1). Should the donkey try to move, elevate the chin and bend the donkey's neck round your body, so that his head is extended and at right angles or more to the line of his body.

Less friendly donkeys must be manoeuvred into a corner situation before being caught. One or two people should gently drive the donkey into the corner of a field or yard. When the donkey feels totally surrounded he will 'freeze' in a stance of submission, head slightly lowered, tail down, and hind feet well under the body. He can then be approached slowly along his side. The handler then slides one arm over his neck, the other under the head, to achieve the hold shown in Fig. 9.1. Once caught, donkeys may be secured by a head collar or halter.

Head collars and halters

Applying a head collar is the same as for a horse. The shape of the head collar does, however, need to be different. A donkey's head in relation to its body is relatively bigger than a horse's. Thus you will need a 'cob' or even a full 'horse' size head collar for an average 10.2 hands donkey. The noseband of such a head collar will then seem too large, so always purchase head collars for donkeys that have *adjustable* nose bands (Fig. 9.2).

Halters, with their infinite variety of shape, might seem a better alternative, but there is one draw-back. Many donkeys are very sensitive about anything touching their ears. Donkeys which are otherwise tractable may pull away when a halter is put over their ears. To avoid touching the ears, unthread the chin-strap element, make a big loop of the nose-band/poll-strap elements, drop them over the ears without touching, and then rethread the chin-strap and adjust. Having caught and haltered the donkey, it is now possible to lead it.

Leading

Donkeys can be taught to lead. They are creatures of habit, so gentle repetition is the key. The best time to start (as with many other procedures) is when the donkey is a foal. The younger animal is more receptive to learning new regimes. The foal will follow its mother instinctively, so if one handler leads the mother and another handler leads the foal behind the mother, the process becomes automatic.

Teaching a single donkey to lead involves urging rather than pulling (Fig. 9.3). The handler will need to stand initially behind the donkey's left

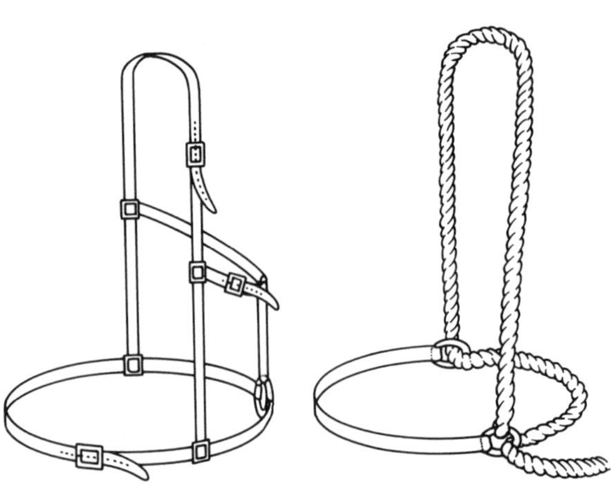

FIG. 9.2 A head collar and halter suitable for donkeys. The head collar (*left*) is adjustable at three buckles; the halter is infinitely adjustable.

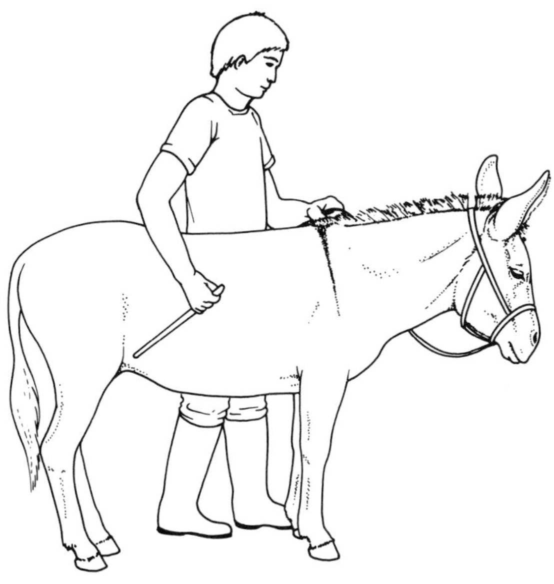

FIG. 9.3 Teaching a donkey to lead involves urging rather than pulling.

shoulder, holding the lead rein in the *left* hand. The right hand holds a tap stick, as a guide. Easing the lead rein and squeezing the tap against the donkey's right side flank will urge the donkey forward. Speed will be determined by the pressure of the stick and how far back towards the hind quarters the handler elects to remain.

All physical instructions (or 'aids') to the donkey should be reinforced by voice commands (e.g. stand!, trot!, etc.), particularly if the donkey is to progress to training for driving.

The process should be repeated at least once a day for 15 minutes. As the donkey accepts the situation, so the handler can adopt a more forward (relative to the donkey) position. Eventually the donkey will be leadable with the handler's shoulder parallel to the donkey's jowl, then the right hand may be transferred to the lead rope.

Saddlery

Many donkeys in North Africa, e.g. the Sudan, can be guided with soft taps to the side of the neck. Many European donkeys are steered by reins to the head collar. However, in the relatively confined and heavily-trafficked spaces of the United Kingdom, it is thought prudent to use a bit in the donkey's mouth. It need only be of the mildest type, e.g. a vulcanite bar snaffle, but it is wise to have large rings or cheek pieces to stop the bit sliding through the mouth.

Fitting a saddle to a donkey is a special art. Suffice it to say that the donkey's back is a very different shape from a pony's. Donkey saddles, therefore, need to be a special fit. Such a saddle will not fit most ponies. Most donkeys will need a crupper (from rear of saddle to base of tail) to stop the saddle sliding forward. Harness and rugs also need to be manufactured to the shape of the donkey. Consult your local saddler.

Long Reining

The average adult (10.2 hands or 105 cm high) can support a rider or pack weight up to a maximum of 112 lb (8 stone or 50 kg approx.). Many adult people are, therefore, too heavy to ride a donkey. Alternatively, since a donkey can pull on level land nearly three times the weight that he can carry, adult people can use donkeys for private driving. The vehicles used should be as light as possible, consistent with safety. Frames and shafts should be aluminium, and wheels can be of the bicycle or light motorcycle type. In more hilly areas, pairs, tandems or teams of donkeys can supply the extra power. The vehicles should then be four-wheeled and fitted with brakes. Guidance on training for harness work can be obtained from books or by contacting the driving division of the Donkey Breed Society.

The basic training for harness work is, however, 'long reining' (Fig. 9.4). This has three main aims:

1. To accustom the donkey to commands by voice and rein from a distance.
2. To accustom the donkey to the paraphernalia of harness, particularly reins and traces around the hind legs. Ultimately the sensations and exertions of draught can be introduced.
3. To accustom the donkey to proceeding in the face of traffic, crowds and other distractions, in *advance* of the handler.

Again it should be remembered that donkeys are well adapted to work at the slower gaits. Harness pace for donkeys should not exceed the 'medium' trot.

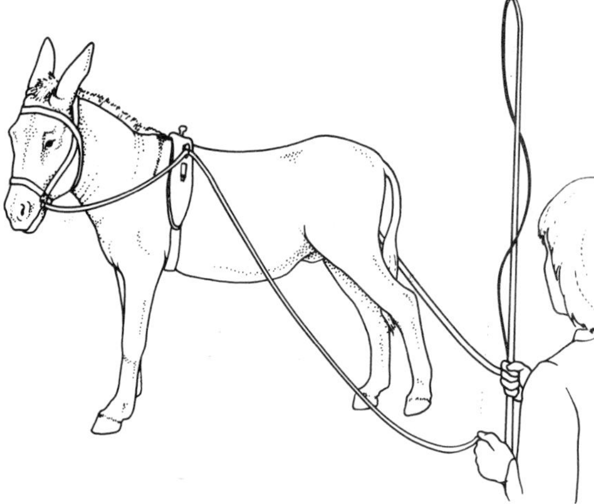

Fig. 9.4 The basic training for harness work is 'long reining'.

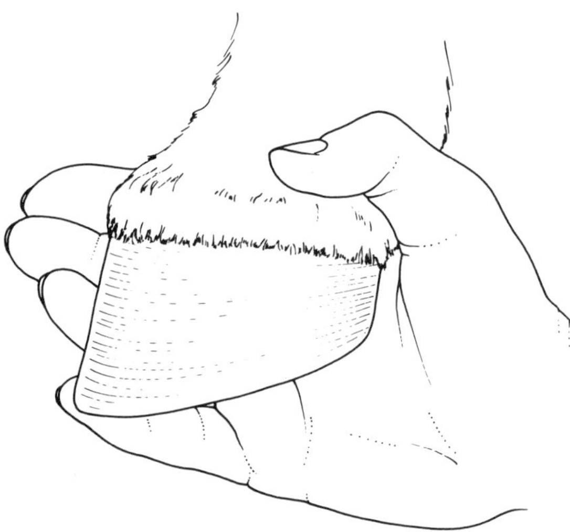

Fig. 9.5 The relation between the size of a donkey's foot and a man's hand. Pressure on the coronary band is painful and should be avoided.

Handling in Special Situations

Foot Care

'*A comfortable patient is more likely to be a co-operative patient*' (Tom Williams, Hereford School of Farriery). It is relatively easy to make a donkey uncomfortable during routine foot care.

Firstly, because of the relatively small size of a donkey's foot, compared with either a horse's foot or a man's hand (Fig. 9.5), the handler can inadvertently apply pressure to the coronary band — a sensitive structure. Donkey handlers should, therefore, school themselves to grip the donkey's foot with *only* those fingers which rest directly on the hoof, and not those which overflow on to coronary band or skin.

Secondly, there is a great temptation when using the farrier's leg grip, to elevate the donkey's foot above the handler's knee, as for a horse. Because of the donkey's lack of height, this applies an intolerable twist to the leg. Handlers must school themselves, for the sake of the donkey's comfort, to grip the donkey's leg between their calf-muscles (i.e. below their knees) (Fig. 9.6). Adult donkeys, who have not previously had regular foot attention, may vigorously resent attempts to lift their hind legs for foot care. This may place the handler at risk of injury.

If the handlers are certain they cannot adequately restrain the donkey's hind leg by hands alone, a modified 'army side-line' may be useful (Fig. 9.7). A non-slip bowline loop is made in one end of a cotton rope and placed over the donkey's neck. The handler moves the donkey forward so that the main bight of the rope trails between the donkey's hind legs. The farrier, at a safe distance picks up the bight around the required leg, passes it twice around itself between leg and shoulder and gives the tail of the rope to the handler. By gently shortening this device the hind foot is elevated.

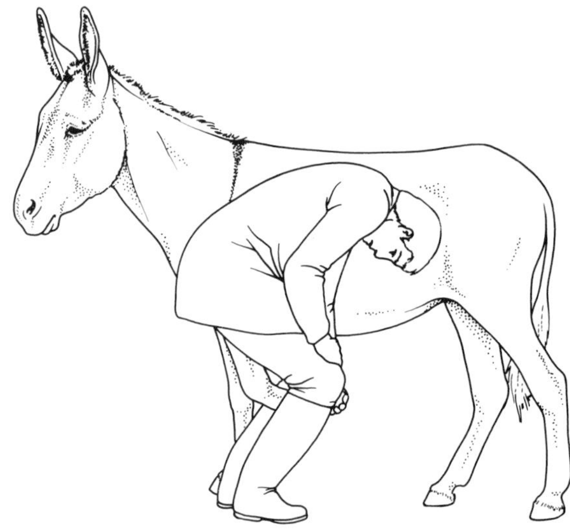

Fig. 9.6 Farrier's stance for a donkey. Note that the foot is held up *below* the operator's knee.

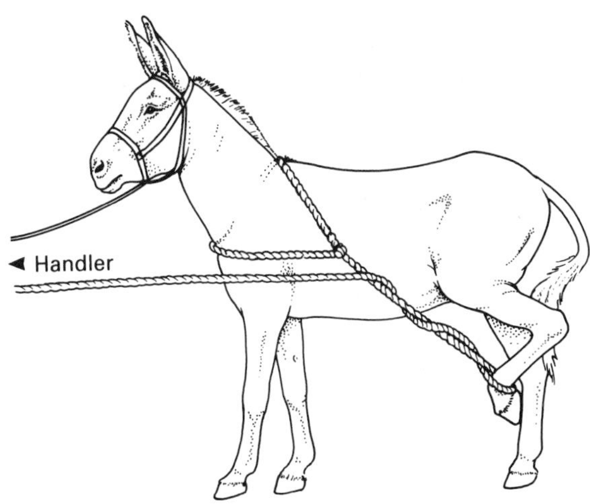

FIG. 9.7 Army side-line to secure the hind foot of a fractious donkey for farriery.

Most donkeys quickly accept the inevitable. If this procedure is done quietly, then by the second or third attempt the donkey will allow its foot to be picked up without this device. If, however, the procedure causes the donkey to panic, then chemical restraint (sedatives administered by a veterinarian) should be employed.

Tooth Care

Severe restraint for such techniques as tooth rasping and descaling tartar is counter-productive. The preferred method is the 'chin hold' (Fig. 9.8).

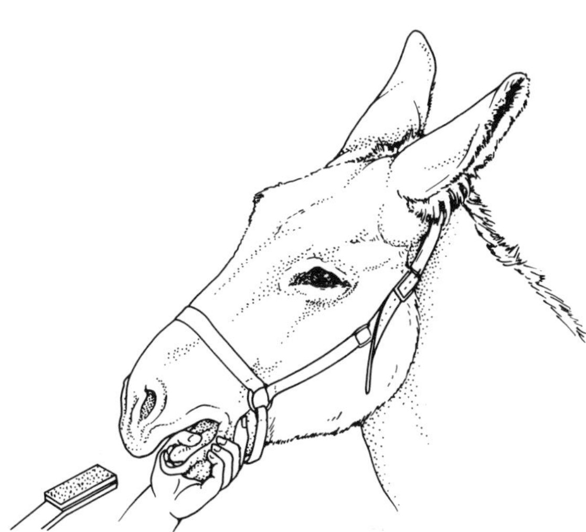

FIG. 9.8 The 'chin hold' for tooth care.

The *right handed* operator should place the donkey's chin in his left hand palm. Curl the fingers around the donkey's left lower ramus, and insert the thumb across the bars of the mouth, posterior to the canine teeth (tushes), approximately where the bit of a bridle would go. The donkey finds it difficult to pull against this grip, and it persuades him to hold his mouth open, thus facilitating the introduction of a tooth rasp or torch-light inspection. If the donkey backs away, do *not* pull forward on his head, but manoeuvre his hind quarters into a corner of the enclosure or stable.

Emergency Restraint

The 'chin hold', with the handler's other hand holding the base of an ear, can be employed in exceptional circumstances to restrain a struggling or panicking donkey, for example, to restrain a donkey caught in barbed wire whilst attempts are made to untangle it. This hold should only be employed for the minimum time necessary and should be replaced by chemical restraint from a veterinarian as soon as possible.

The 'twitch' (a loop of fine rope applied to the upper lip and twisted tight) is, in the author's experience, ineffective and counter-productive in that it only seems to frighten donkeys.

Loading

By playing on their inquisitive nature and coaxing with items of food, donkeys can usually be persuaded to enter weight crates, transporters, stocks, etc., without fear. By repeating such procedures regularly, they can be trained to enter such structures on command.

When an untrained animal has to be loaded and time does not permit coaxing or training, forced loading may be necessary. This can be accomplished without harm to the animal, by employing the method commonly seen on race-courses to entice reluctant thoroughbreds into starting-stalls (Fig. 9.9). Two handlers stand on either side of the donkey and hold a loop of rope (made perhaps from a spare lead rope) passed around the donkey's rear. This loop is used to propel the animal forward. The two handlers can apply pressure to the donkey's shoulders to guide its direction. A

Fig. 9.9 Leading a reluctant donkey.

third handler precedes the donkey whilst holding a lead rope from the head collar. No *pull* should be applied to this rope other than that which is necessary for directional guidance.

Chemical sedation, administered by a veterinarian, may make some of these procedures easier, but when administered prior to travelling it may impair the donkey's ability to balance and thus be contra-indicated.

Mules

Mules and hinnies present the handler with an interesting mixture of the attributes of donkeys and horses. They are, in general, less phlegmatic than donkeys, more easily startled and considerably more athletic. The training of these animals is closer to that of horses, and may well take longer than donkeys. Once the mule has accepted his training, however, his capacity and enthusiasm for work is astounding.

Mules exhibit great hardiness and surefootedness and a degree of stubborness. These characteristics can make them difficult pupils if training does not start at an early age (2–3 years old). In particular, it is wise to accustom mules and hinnies to having their feet handled from foal age onwards.

The reader may wish to consult the British Mule Society for specialist advice.

Further Reading

Ellis, R., Ellis, V. and Claxton, J. (1990) *Donkey Driving*. J. A. Allen, c/o The British Driving Society, Royal Mews, London.

French, J. M. (1988) Provision of cubicles for housed donkeys. In *Proc. Int. Cong. on Applied Ethology in Farm Animals*, Editors Unshelm, J., Van Putten, G., Zeeb, K. and Ekesbo, I. pp 287–290.

Moehlman, P. D. R. (1974) *Behaviour and Ecology of Feral Asses* (*Equus asinus*). PhD thesis, Univ. Wisconsin, Madison.

Svendsen, E. D. (ed.) (1989) *The Professional Handbook of the Donkey*. The Donkey Sanctuary, Sidmouth, Devon EX10 0NU.

Walrond, S. (1988) *A Guide to Driving Horses*. Pelham Books Ltd, London.

Yousef, M. K. (1985) *Stress Physiology in Livestock*. Volumes I and II. CRC Press Inc, Florida, US.

Useful Addresses

Donkey Breed Society Driving Section
Mrs C. Pinnegar
Gambles Lane
Woodmancote
Cheltenham, Glos. GL52 4PU

British Mule Society
Mrs L. V. Travis
Hope Mount Farm
Top of Hope
Alstonfield
Ashbourne, Derby DE6 2FR

The Donkey Sanctuary
Sidmouth
Devon EX10 0NU

10

Cats

JOSEPHINE WILLS

Introduction

Many cats can be difficult to handle. They are agile and quick and their independent nature inevitably means that they may tolerate handling for only short periods. In addition to sharp teeth, cats also have four sets of claws that the handler needs to consider. The degree of restraint exercised is very dependent on the individual cat. Most cats respond best to a loose and gentle approach rather than forceful restraint. Much can be achieved by an unhurried but firm technique.

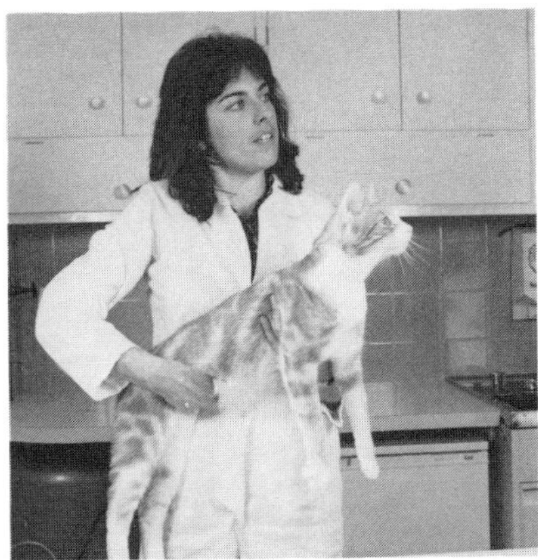

FIG. 10.1 Lifting a cat, with one hand supporting the sternum and the other supporting the abdomen.

Handling

A cat should be approached in a calm but confident manner, speaking to it quietly. When a stranger tries to remove a frightened cat from a cage or basket, the cat may well adopt an instinctive defensive posture. However, unless the cat warns the handler either by hissing or growling, it is usually safe to put a hand on top of the cat's head. Stroking the top of a cat's head and running the hand confidently alongs its back will normally reassure it. The cat can then be lifted up.

Cats can be lifted in a number of ways, depending on their age and body weight, whether the cat is passive or fractious, whether it is in a cage, basket or on the floor, and the individual preference of the handler. The cat may be lifted up by passing one hand over the chest wall and supporting the sternum, whilst the other hand supports the abdomen from the other side (Fig. 10.1). Once picked up, the cat should be held firmly towards the handler's body (Fig. 10.2). This method is suitable for a relaxed cat. If it is not possible to pick a cat up directly from above, for example when it is in a cage or basket, it can be lifted by grasping both forelegs from the elbows and gently pulling it free from the cage. It can then be tucked under one arm, with the hand of that arm holding the forelegs and with the second hand stroking the cat's head (Fig. 10.3) or gently 'scruffing' it (Fig. 10.4). Kittens can be handled similarly, but those lifted or carried gently by their scruff will often adopt a very relaxed, submissive posture. Cats of dubious temperament can be lifted up by the scruff, with the

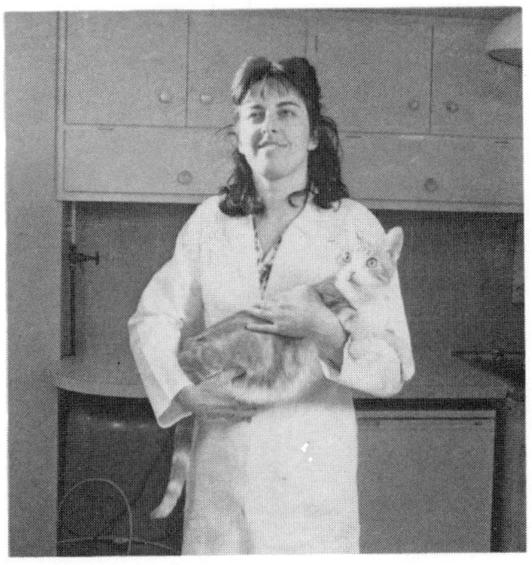

FIG. 10.2 Holding a cat firmly.

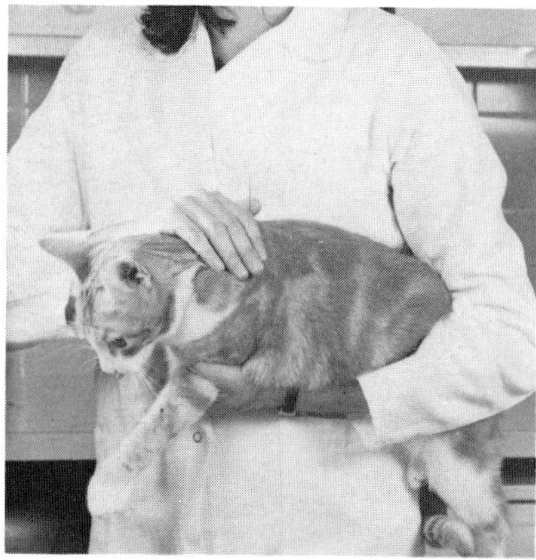

FIG. 10.3 A cat tucked under one arm, with the other hand stroking its head.

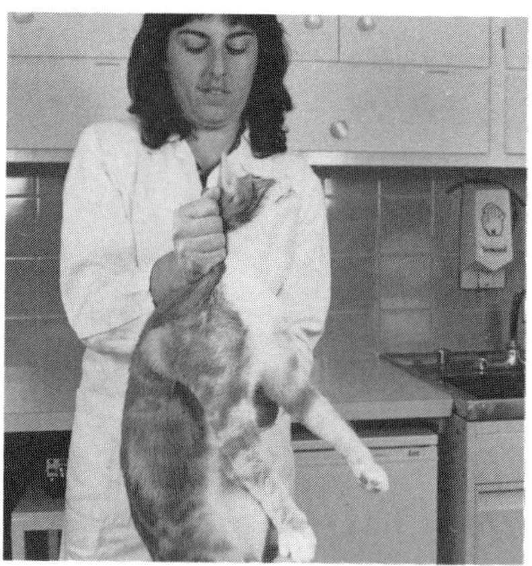

FIG. 10.5 Lifting a cat by the scruff.

legs pointing away from the handler and the second hand supporting the cat at the sternum (Fig. 10.5) or taking the weight on the hind quarters.

Holding for General Examination

As before, techniques that can be used for holding a cat for general examination depend on whether the cat is amenable to handling, whether the examiner is single handed, the area to be examined, the nature and extent of the treatment or sampling required, and individual preference. In general, cats respond best to light restraint. However, if the handler is single-handed, firmer restraint may be necessary by, for example, scruffing the cat with one hand whilst leaving the other free to examine the body. If assistance is available, the cat can be presented to the examiner in different ways depending on the site to be examined. For examination of the head, an assistant holds the cat's forelegs to protect the examiner from the front claws, facing the cat towards the examiner (Fig. 10.6). The body of the cat can be restrained in the assistant's arms. This leaves the examiner able to hold the cat's head or scruff with one hand, and have the other hand free for examination, sampling or dosing. For examination of the rest of the body, passive cats can be restrained by gently holding the cat's shoulders and forelegs with both hands with the cat facing the assistant (Fig. 10.7).

FIG. 10.4 The neck may be gently 'scruffed'.

FIG. 10.6 Holding for examination of the head.

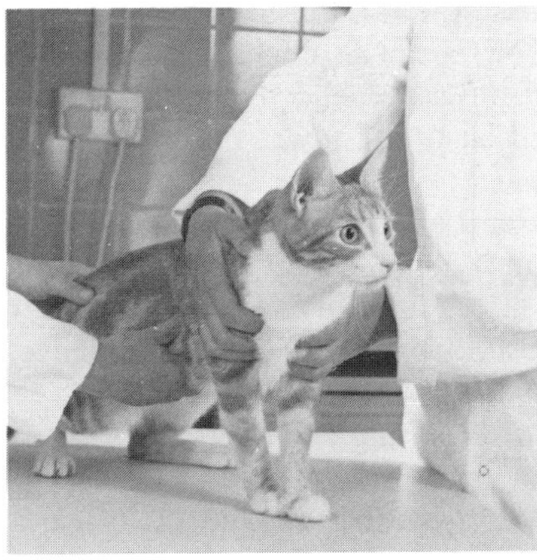

FIG. 10.7 Holding for examination of the body.

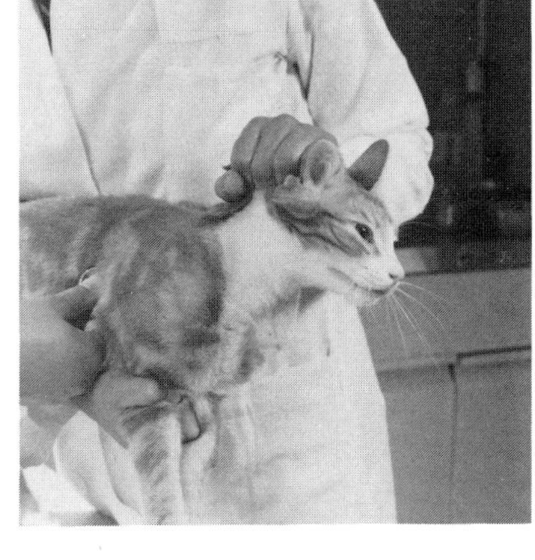

FIG. 10.9 Restraining a restless cat by using the scruff.

If the cat becomes restless or less amenable, the grip can be changed to be more firm by either holding the head under the jaw or scruffing it with one hand, whilst the other hand holds the cat's forelegs and restrains it against the assistant's body (Figs 10.8 and 10.9). If the cat is particularly difficult, it can be held in lateral recumbency, with one hand gripping the scruff, whilst the other hand grasps the hind legs (Fig. 10.10). It is best to use only as much restraint as necessary, provided that firmer control can be applied rapidly.

Temporary immobilization of some cats has been achieved by some workers by placing an elastic band around the base of both the cat's ears (Leedy *et al*, 1983). The cat then usually assumes a semi-crouched position and remains relatively motionless for a period. In the author's experience, this method is unreliable and should never be used for more than a few minutes.

Restraint of a Fractious Cat

It may not be possible to restrain a fractious cat safely just by scruffing it and grasping the hind legs. Some cats are capable of letting their neck sink into their shoulders so that the 'scruff' disappears. Also, adult tom cats often have very thickened scruffs which can be difficult to grasp firmly for any length of time. In these cases, it is often advisable to resort to using a large, thick towel to wrap the cat in. For examination of the head, the whole

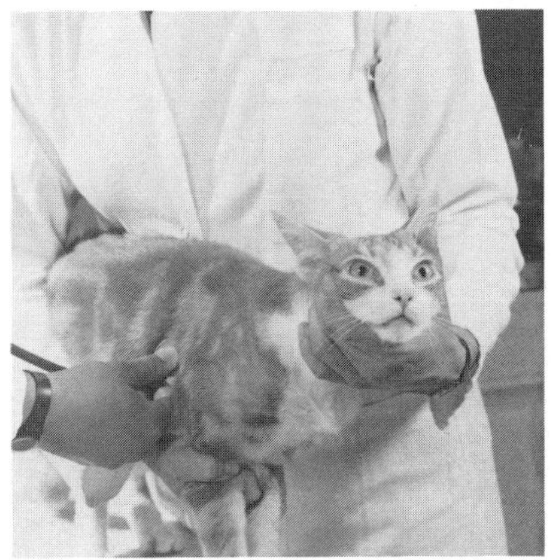

FIG. 10.8 Restraining a restless cat by holding the jaw.

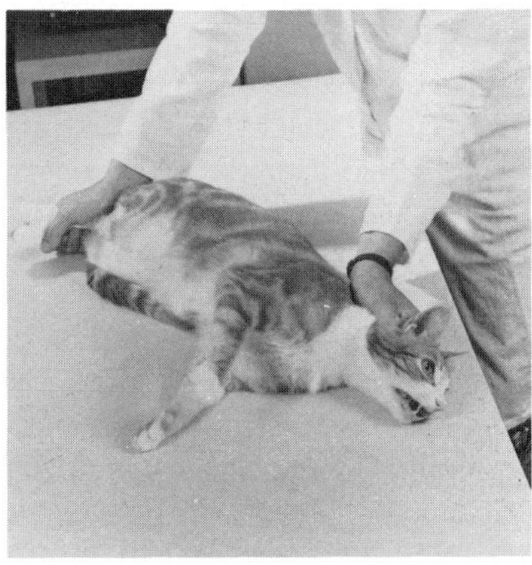

FIG. 10.10 Holding a difficult cat in lateral recumbency.

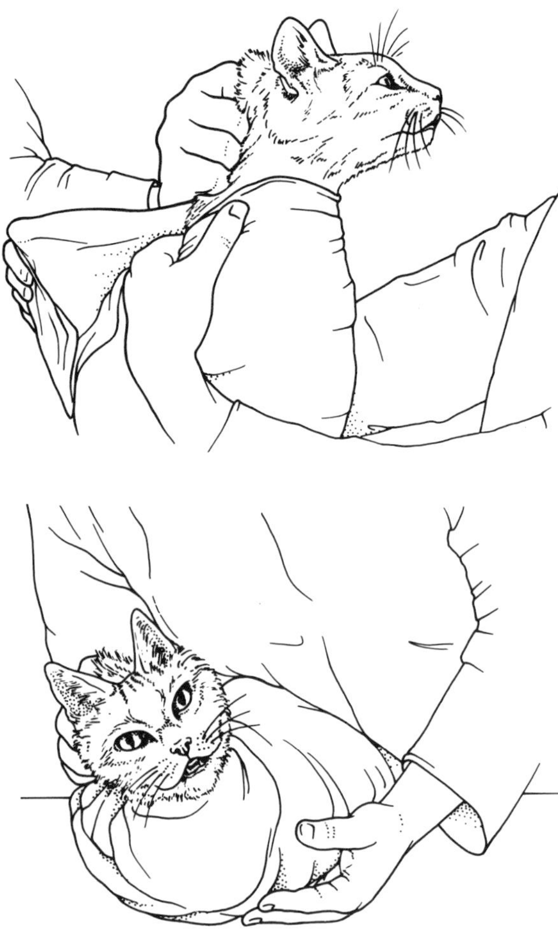

FIG. 10.11 Securely wrapping a cat in a towel.

body can be firmly wrapped in a towel, making sure the forelegs and hind legs are secure (Fig. 10.11). The assistant can then hold the enveloped cat firmly, leaving the examiner free to control the head. It is best not to let the cat have sight of the towel or blanket beforehand. This is a very useful method of restraint which owners can learn to use. It can protect them from being scratched; it can also be used when any medication has to be given at home, such as the administration of ear or eye medication, liquid medicines or tablets.

The wrapping-up-in-a-towel method has been used in the design of restraining bags. These bags are usually made form strong nylon material and are waterproof. The cat is placed in the bag which is then zipped up, leaving only the head showing. There are four leg holes which can be unzipped to allow exposure of a leg for treatment (Fig. 10.12). These bags are also used for cephalic venepuncture. Towels can also be utilized to pack the area around the cat's head so that any biting can be directed on to the towel and not the assistant or examiner.

Restraint of a vicious cat can be a considerable problem. Even if the cat has been grasped by the scruff successfully, it may contort itself and succeed in striking the handler with its hind legs. A cat which behaves in this way can often be restrained by crowding it into a corner with a thick towel or blanket, then grasping it through the protective layer. Alternatively, a thick pair of gardening or animal gauntlets can be used to protect the handler's hands, but the arms must be well protected as well. The cat's claws must also be kept away from people's faces.

Catching and Handling Feral Cats

Occasionally, it may be necessary to catch a feral cat that is injured or as part of a control scheme for neutering, identification and treatment. Feral cats are usually extremely difficult to handle, despite the recommendations set out above. They are very frightened, not used to man or to any quick or unexpected movements. The most satisfactory method of capturing a feral cat is to use a cat trap with a squeeze-back facility (UFAW, 1981). The captured cat is then pressed against the mesh on the side of the cage, so that an injection of an immobilising agent can be given, and then the cat may be handled safely.

If the cat is presented in a conventional cat basket, a thick blanket can be pushed into the cage, through a narrowly opened lid, to press the

FIG. 10.12 A cat in a restraining bag, with foreleg extended.

cat against the mesh for the injection to be given. If injection techniques are not possible due to the design of the cat basket or carrier, then it will be necessary to use gaseous anaesthesia to immobilize the cat by placing the carrier in a suitable transport polythene bag, and introducing the gaseous anaesthetic into the bag. The cat can be removed as soon as anaesthesia has reached a level sufficient to immobilize it.

Occasionally it may be necessary to catch a feral cat without the aid of a cat trap, e.g. a feral cat which has escaped from its container and is loose in the veterinary hospital. The cat will often seek refuge under a cupboard. Once the cat is cornered, a set of cat-catchers or cat tongs can be used to catch and restrain it. When used correctly they do not cause injury to the cat.

Specific Restraint Techniques

Most of the procedures that involve giving medication or taking a sample are best carried out with two people. One person restrains the cat in a way in which the medication can be given accurately and quickly, providing full protection for themselves and the second person who is giving the medication. In certain circumstances, procedures may have to be carried out single-handedly, in which case the restraint procedure has to be adapted so that one person can restrain the cat, carry out the procedure quickly and effectively, whilst protecting themselves from the teeth and claws of an uncooperative cat.

Taking the Rectal Temperature

Clinical examination of a cat often involves taking the rectal temperature with a thermometer. If assistance is available, then the cat can be presented to the clinician who can then grip the base of the cat's tail with one hand, to control movement of the cat's hind quarters, and use the other hand to insert the rectal thermometer. If single-handed, the handler can restrain the cat by holding it towards the body with the elbow, using the hand of the same arm to grip the base of the tail. This leaves a free hand to take the temperature (Fig. 10.13).

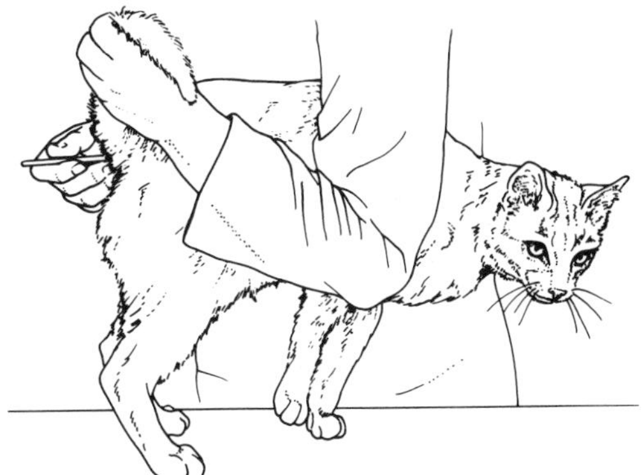

Fig. 10.13 Taking a cat's rectal temperature.

Liquid Feeding or Medication

Liquid medication, such as an anthelmintic, anti-diarrhoeal agent or antibiotic, is best administered using a plastic syringe. If assistance is available, then a cat can be presented to the clinician as in Fig. 10.6. The assistant presents the cat to the clinician whilst holding the cat's forelegs to prevent it from scratching. The clinician can then restrain the cat's head with one hand, and introduce the end of the syringe through the side of the cat's mouth. If single-handed, the most comfortable way of administering medication or liquid food is to sit down with the cat in sternal recumbency on a towel on the lap, facing away from the handler. One hand can be used to grip the cat's upper jaw and bend the head backwards slightly, whilst the other hand is used to introduce the food into the side of the cat's mouth gradually, via the syringe (Fig. 10.14). If the cat becomes fractious, further control can be obtained by pressing in on the cat with the elbows and chest. The towel can be used to wrap around the cat's forelegs if it tries to claw the syringe. Should the cat become too difficult to handle with these minimal restraint procedures, then it can be wrapped fully in the towel, as in Fig. 10.11, before the medication is given.

Giving a Tablet

If an assistant is available, then the cat can be restrained by the forelegs, as described previously. The cat's head is then gripped with one hand, with the thumb and forefinger at the angle of the cat's jaw. The head is tilted well back and pressure at

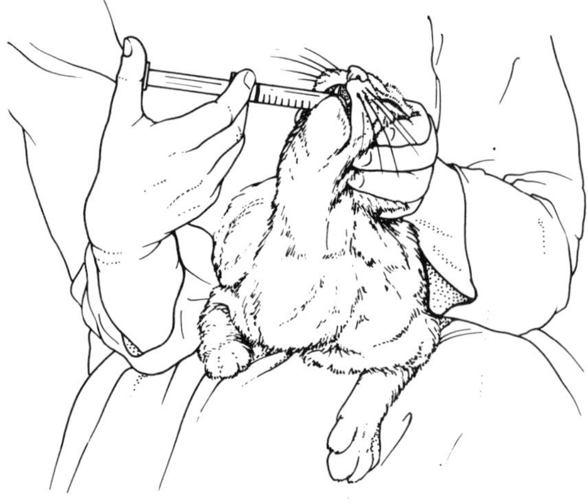

Fig. 10.14 Giving a liquid diet via a syringe, single-handed.

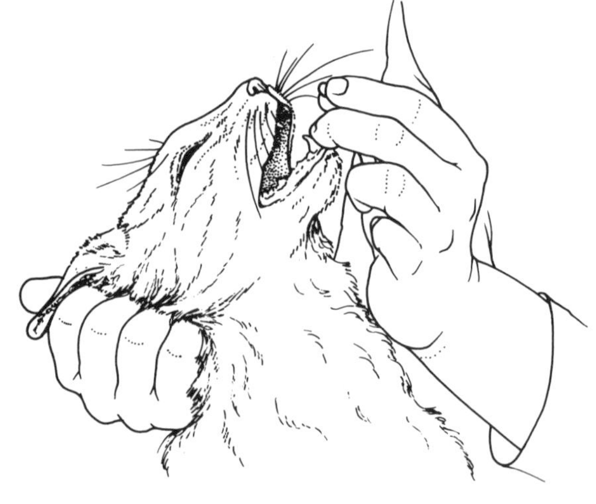

Fig. 10.16 Giving a tablet single-handed

the angle of the jaw will ease the mouth open. With the other hand, the tablet is held with the thumb and forefinger and the second and third fingers are used to press down on the lower jaw (Fig. 10.15). The tablet can then be placed or dropped as far back on the tongue as possible. Usually, a tablet placed far enough on the back of the tongue will induce a swallowing reflex but, if this does not occur, the cat's mouth is closed and held closed. Keeping the cat's head tilted upwards, the cat's throat is gently stroked until it swallows.

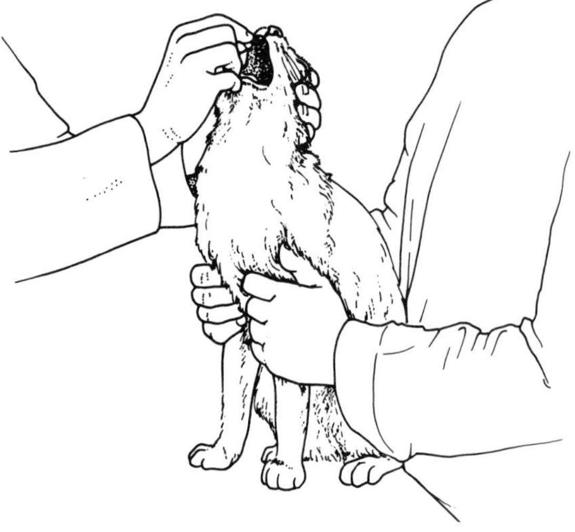

Fig. 10.15 Giving a tablet.

A tablet can be given successfully even if the handler is single-handed. The cat is sat on a table facing the handler. The head is held with one hand with the thumb and fingers gripping the cat's ear and scruff (Fig. 10.16). The cat's head is then twisted until the nose points upwards. The mouth can then be opened as before and the tablet given. If there is a serious danger of being bitten, a tablet doser can be used. This can be made out of a plastic syringe, with the end cut off to allow the tablet to fit into the barrel. The plunger is retained to 'fire' the tablet at the back of the throat.

Giving Ear Medication

If assistance is available, then the cat can be restrained by the forelegs facing the handler as described previously. With the forefinger and thumb of one hand, the pinna of the affected ear can be held so that the head can be gently twisted so that it faces upwards (Fig. 10.17). For more control, the scruff can be incorporated in the hand being used. The medication can then be applied with the other hand.

Ear medication can be administered by one person, by lying the cat on its sternum, facing away from the handler. Grasping the cat's scruff with one hand, the head is tilted so that the affected ear is uppermost. The forearm and elbow can be used to give additional restraint, pressing the cat towards the body (Fig. 10.18). The medication is given with the other hand.

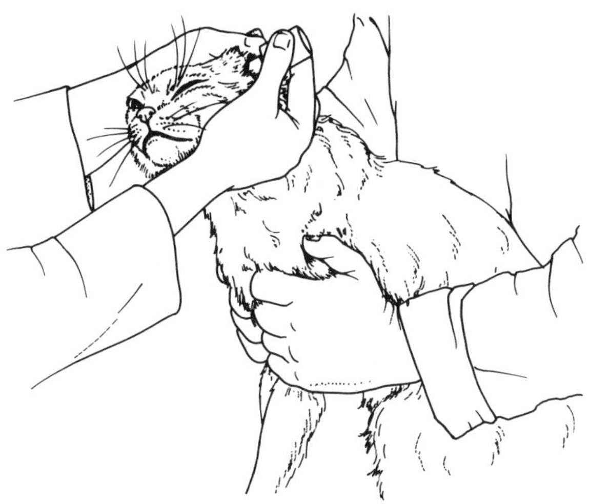

Fig. 10.17 Applying ear medication.

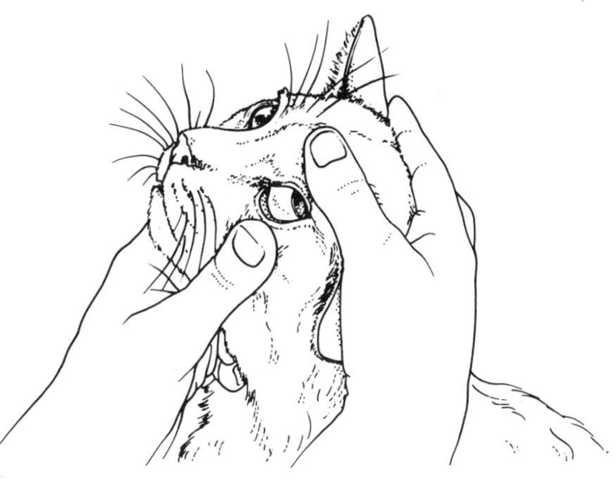

Fig. 10.19 Examination of the conjunctivae.

Giving Eye Medication

Examination of the eye and conjunctiva is best achieved with an assistant facing the cat towards the handler and restraining the cat by its forelegs and elbows. The conjunctival epithelium can be examined by cupping the cat's head in both hands, and using the thumb of one hand gently to pull down the lower lid so that it is everted. Meanwhile the thumb of the other hand applies gentle pressure to the upper lid (Fig. 10.19). This procedure results in the third eyelid coming across the eye.

Eye medication can be given when there is no assistance available. The cat is sat on the table facing away from the handler. Its body is tucked into the handler's and restrained by the forearms and elbows (Fig. 10.20). The cat's head is restrained in one hand, with the thumb pressing on the top eyelid to keep it open. With the remaining hand, the third finger is used to open the lower lid, whilst the medication is given with the thumb and forefinger.

Subcutaneous Injection

This procedure is best carried out with no assistance. The cat is placed in sternal recumbency on the table, facing away. With one hand grasping the cat by its scruff, the injection is given with the other hand into the raised skin of the scruff (Fig. 10.21). Most cats tolerate this procedure well, but

Fig. 10.18 Applying ear medication single-handed.

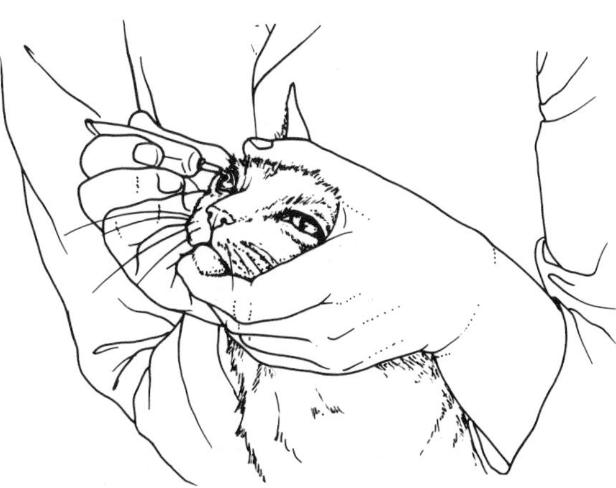

Fig. 10.20 Application of eye medication single-handed.

Fig. 10.21 Subcutaneous injection into the scruff of the neck.

Fig. 10.22 Intramuscular injection.

additional restraint can be achieved with the forearms and elbows and pressing down onto the cat with the chest.

Intramuscular Injection

The safest way to give an intramuscular injection is first to have the cat adequately restrained. The most satisfactory muscle mass for an intramuscular injection is the quadriceps group anterior to the femur. The assistant faces the cat's hind quarters towards the clinician, and grasps the cat's scruff with the one hand. The clinician restrains the hind leg with one hand, and gives the injection with the other (Fig. 10.22). Intramuscular injections can be given to feral or fractious cats, through the mesh of a cage with a squeeze-back facility (see section on catching and handling feral cats).

Intravenous Injection

In cats, the cephalic vein is the most convenient way to given an intravenous injection. The cat is sat on a table, facing the clinician. The assistant holds the cat's head under its chin with one hand and uses both forearms and chest to hold the cat. The cat's foreleg is pushed forward at the elbow with the other hand. The thumb and forefinger are used at the crook of the cat's elbow to apply a tourniquet effect, to raise the cephalic vein (Fig. 10.23). The cat can be restrained further by changing the grip from around the chin to one scruffing and twisting the cat's head away from the clinician's face (Fig. 10.24).

After the needle has entered the vein, the assistant releases the thumb pressure on the vein, so that the intravenous injection can be given, whilst still pushing the cat's foreleg forward at the elbow (Fig. 10.25).

Fig. 10.23 Restraint for raising the cephalic vein.

Fig. 10.24 Firmer restraint for raising the cephalic vein.

Blood Sampling

Blood samples can be taken safely from conscious cats, either from the cephalic or the jugular vein. The cephalic vein is commonly used (Fig. 10.26). The jugular vein can be used if comparatively large volumes of blood are required quickly, e.g. from a cat used as a blood donor, and is better than the cephalic vein in kittens under six weeks old.

The method of restraint for taking a blood sample from the cephalic vein is the same as that described for intravenous injection as illustrated in Figs 10.24 and 10.25. However, the thumb and forefinger remain as a tourniquet throughout the procedure (Fig. 10.26). Wrapping in a towel can be useful for more difficult cats, leaving a foreleg free for sampling (Fig. 10.27).

Fig. 10.25 Releasing thumb pressure, for intravenous injection.

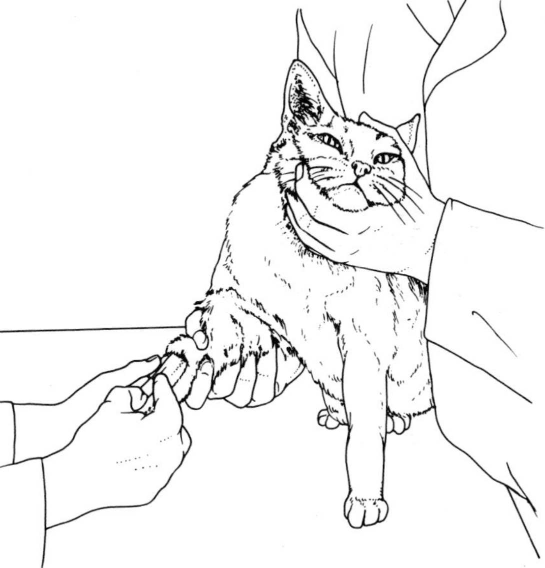

Fig. 10.26 Obtaining a blood sample from the cephalic vein.

Jugular venepuncture is tolerated well by most cats. The cat is restrained in dorsal recumbency and placed on the lap of the assistant who sits on a

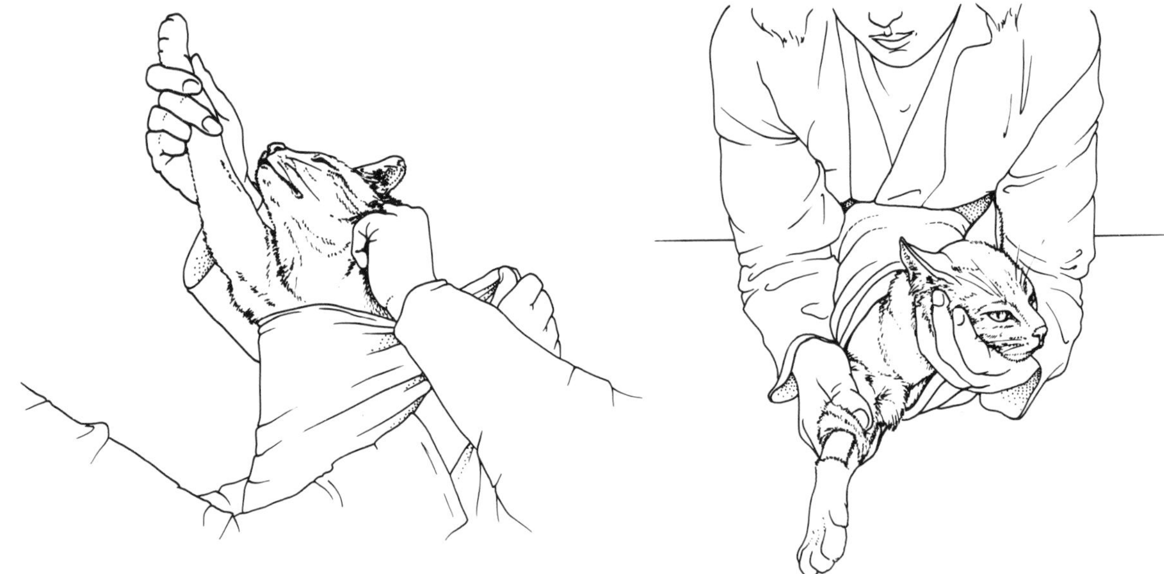

FIG. 10.27 Wrapping a cat in a towel, for intravenous injection or blood sample from the cephalic vein.

chair (Fig. 10.28). If the right jugular vein is used, the assistant holds all four legs in his right hand. The thumb of the assistant's left hand is placed in the jugular furrow and pressure is applied to raise the jugular vein. The clinician restrains the cat's head by its chin with one hand, and uses the other hand to take the blood sample. The positions are reversed for left-handed operators.

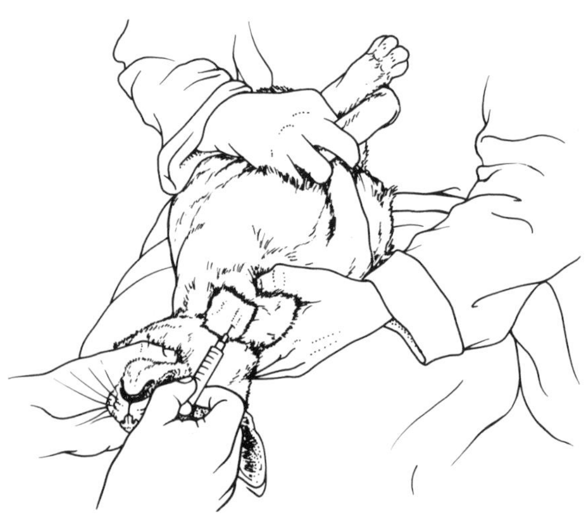

FIG. 10.28 Jugular venepuncture.

Carrying and Transporting Cats

Domestic cats are usually transported from the home to a cattery or hospital in either a cat-carrying basket, cage or box. A wide range of suitable cat-carrying baskets is readily available from the pet trade and veterinarians. Materials used include wickerwork, fibreglass, wood, plastic and plastic-coated wire mesh. Veterinarians often stock cardboard carrying boxes which are reasonably priced, although they have a limited life span.

Before a cat is removed from a basket to a cage and *vice versa*, it is essential to ensure that all doors and windows are closed as well as securing all cat flaps, fireplaces, chimneys and other possible exits, to prevent the cat from escaping if the grip on the cat is relaxed. Cats can be carried within a room as shown in Figs 10.2, 10.3 and 10.4. However, a cat should only be moved from one room to another, or from the veterinary hospital to the owner's car, in a proper basket and not be carried loose. When a cat is to be removed from a basket, the lid should be raised gently and if the cat does not appear to be vicious, a hand is slipped into the basket to restrain the cat before the lid is folded back fully. Similarly, if a cat is placed in a basket, it is put in rear-end first, and one hand is kept firmly on its back while the lid is shut with the other hand.

Recommendations for the transportation of

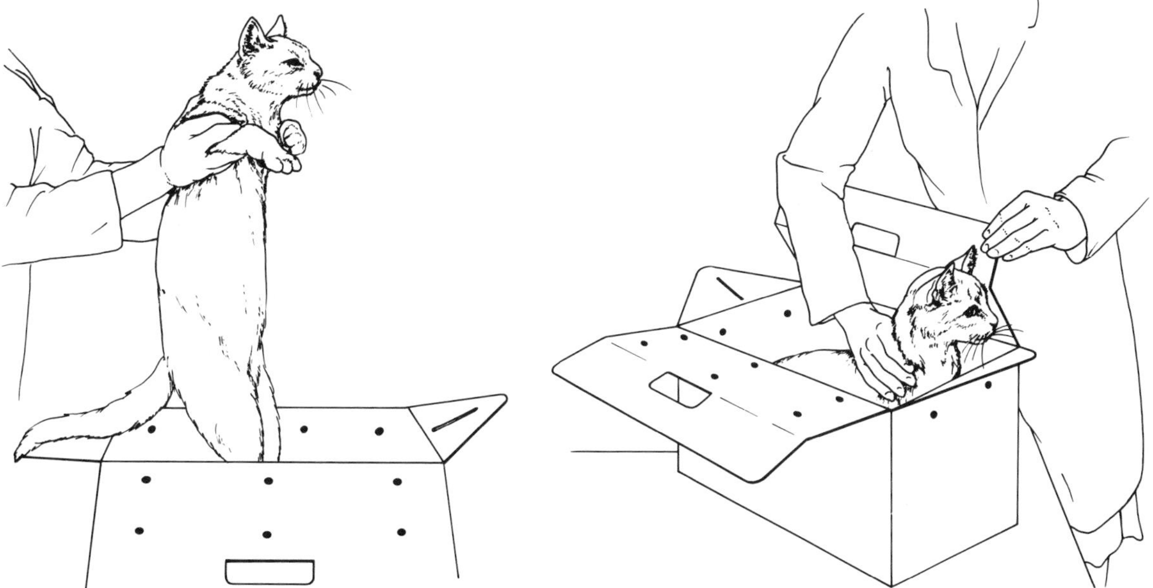

FIG. 10.29 Putting a cat into a cardboard carrying box.

FIG. 10.30 A cat wearing an 'Elizabethan collar' to prevent it from scratching its face and licking areas of the body.

large numbers of cats over medium-range journeys (e.g. within continental Europe) can be found in the UFAW Handbook (UFAW, 1987). Transportation by road is preferable to rail or air. Fibreglass containers (e.g. length 61–68 cm, breadth 36–44 cm, height 40 cm) can be used to accommodate one or two cats. All four sides should have adequate ventilation with holes which cover 10–15 per cent of the surface area. Absorbent bedding should be provided, together with water and food for long journeys.

References

Hurni, H. and Rossback, W. (1987) The laboratory cat, in *The UFAW Handbook on the Care and Management of Laboratory Animals*, Editor Poole T. 6th Edition, pp 476–492.

Leedy, M. G., Fishelson, B. A. and Cooper L. L. (1983) A simple method of restraint for use with cats. *Feline Practice* **13**, 32–33.

Universities Federation for Animal Welfare (UFAW) (1981) Feral cats: notes for veterinary surgeons. *Vet. Rec.* **108**, 301–303.

11

Dogs

TREVOR TURNER

Main Behavioural Characteristics of Dogs

Dogs are social animals. They are happiest when integrated into a family or pack structure. They have the inborn ability to promote, recognize and respond to a whole range of social interactions. The individual members of the pack or family below the pack leader are dominated. They are provided with leadership and direction, and rely on the pack as a whole and its leader in particular, to provide support and act as the decision maker. This is an inherent characteristic of dogs and, although individuals are capable of alteration by experience (learning), it cannot be extinguished. These basic social responses cannot be learned later in life, although experience (training) can reinforce them. The species towards which the dog directs its social responses can, however, be learned. Therefore it is essential that puppies have close contact with people during the first twelve weeks of life. This is most important (even if it only occurs on brief occasions) if the puppy is to become a well adjusted, domestic dog, able to interact with people and accept its ultimate role, irrespective of whether it is a working animal or a family pet. Thus the handling of puppies for a few minutes each day, from the age of approximately eighteen days, helps to establish social interaction with people (imprinting). People are then accepted as 'part of the pack'.

All dogs have genetic potential for both dominant and submissive behaviour. In some of the herding and guarding breeds, the genetic potential for dominance is very strong. In the pack situation, individuals of some of these inherently dominant breeds have to adopt a submissive role, as a result of the learning process which begins with weaning, to fit in with human society.

Within the litter, feeding is highly competitive, and from five weeks of age, play fighting and dominance hierarchies are established. It is from five to twelve weeks of age that man has to be established as the pack leader. This is important even with a one to one relationship between puppy and owner. Puppies which take the lead in aggressive play within the litter or 'family', are likely to establish themselves as dominant animals when adult. If puppies are to accept a subordinate role to man when adult, they should be socialised as early as possible, especially in the case of the more powerful guarding breeds. Unfortunately, this has often been in direct conflict with the need to isolate puppies until immunity to the major canine diseases has been established by vaccination. It may be as late as sixteen to twenty weeks before complete immunity develops.

Social hierarchies within packs of dogs are not usually linear. The pack leader dominates the other dogs (Fig. 11.1) which may be uncertain of their status and whose roles alter according to circumstances and the environment. One, and then another individual, will assume dominance according to the situation. This behaviour can be exploited by a veterinarian who can often control a dog, known to be difficult with the owner, by taking it out of its home environment and examining it in the hospital away from the owner's presence. Owners are often astounded, even disappointed, to find that their dog, far from being unhandleable, can be dealt with gently away from territorial and hierarchical constraints.

In the veterinary context, it is important that no dog should ever be allowed to dominate the humans associated with the pack or family. Otherwise veterinary examination and treatment are likely to be vigorously resisted.

For young dogs, 'play' is an important and

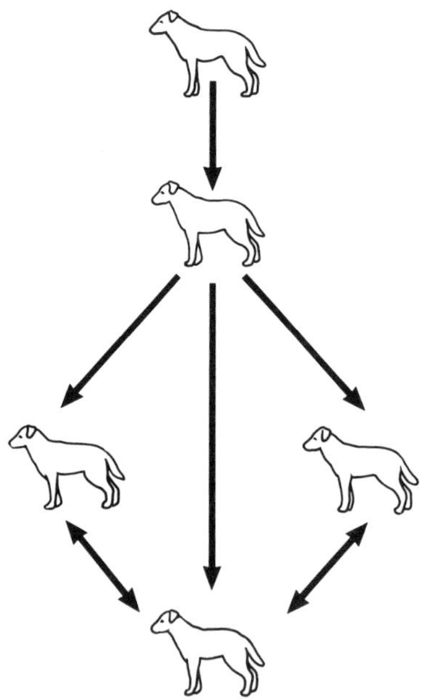

FIG. 11.1 Dominance hierarchy. The pack leader dominates the other dogs.

complex social interaction. It can be used to establish dominance, to deflect aggression or to test out a prospective adversary.

Body Language

'Body language' is complex in the dog and can signify much more than dominance and submission, although these are the most important. Dominance is usually established by body language, although it can be determined by physical aggression and serious fighting within the pack. The dominant dog will meet the adversary's gaze. This can be the owner, another dog, cat or other animal. He stands with the ears forward and the hair on the neck and midline back (hackles) raised, the lips are pulled into a characteristic snarl or the dog may growl and attempt to bite (Fig. 11.2).

Dominance and Submission

Dominance may be expressed to another dog by standing over it, often at right angles. Submission by the dog is characterized by lack of eye contact. The submissive animal crouches down with the ears back and tail lowered. It may roll on its side with one leg raised, exposing its inguinal region (Fig. 11.3). Submissive animals never approach directly but always from the side. Such a dog will try to lick the face or hand of the dominant dog or person.

It is essential that veterinary personnel are able to 'read' and interpret these basic signs in order to handle the animals that are presented every day. In the eyes of the owner, anyone working within the veterinary premises is a 'professional' and will be expected to be able to handle their animal, even if they are incapable of doing so themselves. In the veterinary environment, the normally dominant animal is often in conflict. Thus it may project conflicting signs, for example, growling with the ears back or crouching submissively. The fact that it is giving submissive signals does not mean that it will not bite. Dogs which give contradictory signals are often the most dangerous because of their unpredictability.

If a dog is to allow any examination or veterinary procedure to be carred out, it must be submissive. Before attempting to establish dominance over a dog for these purposes, it is important to learn something about the individual's normal behaviour characteristics as far as this is possible. The owner can be tactfully asked if the dog is aggressive in any situations in the home and if its normal behaviour patterns are predictable or not. It is always better to ask if the dog is friendly rather than 'does it bite?'.

FIG. 11.2 The dominant dog stands with ears forward and hackles raised.

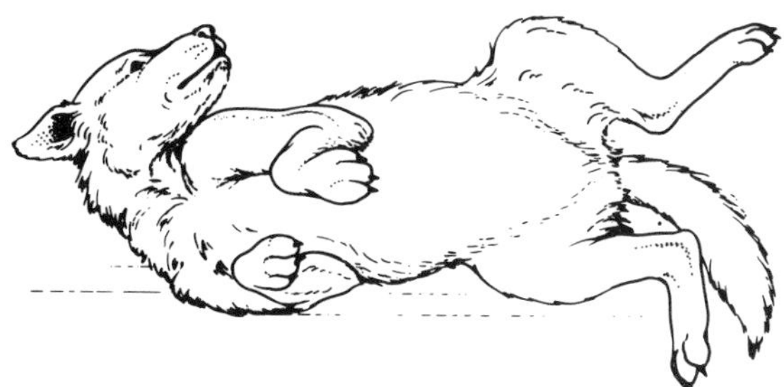

FIG. 11.3 A submissive animal often rolls on to its side.

Unlike man, dominance is part of a dog's instinctive repertoire. Behaving dominantly towards a dog is therefore a learned skill. Those humans who are brought up in an environment with dogs often acquire this skill at an early age. In the home the most effective method of asserting dominance is to pay attention to the dog only when it obeys a command. If it is ignored at all other times, particularly if it is initiating some social activity, it can be manipulated to advantage. Unfortunately, this method is of little practical use to the veterinarian as it is slow, and learning is cumulative. It is often easier for someone who does not live on intimate terms with a particular dog to dominate it. Meeting the animal's gaze and standing tall over it are two ways of expressing dominance. Whilst doing this it is important to observe the dog's behaviour closely. Assess if it is showing any submissive signals or is likely to attack. If the latter is likely, the veterinarian may check the dog and possibly get away with it (i.e. dominate it) whereas, if the owner did this he may be badly bitten; the dog making a bid for dominance.

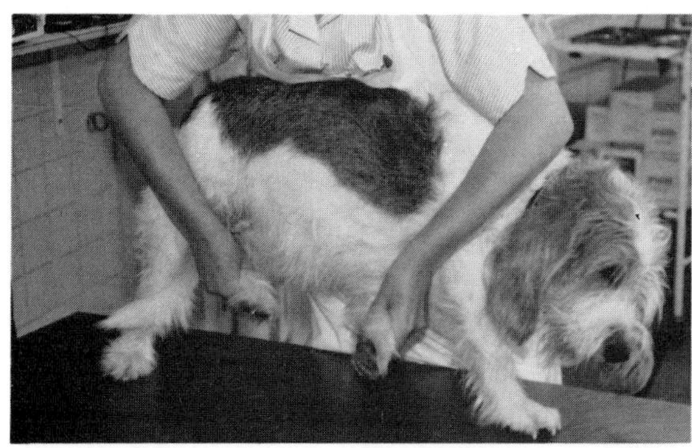

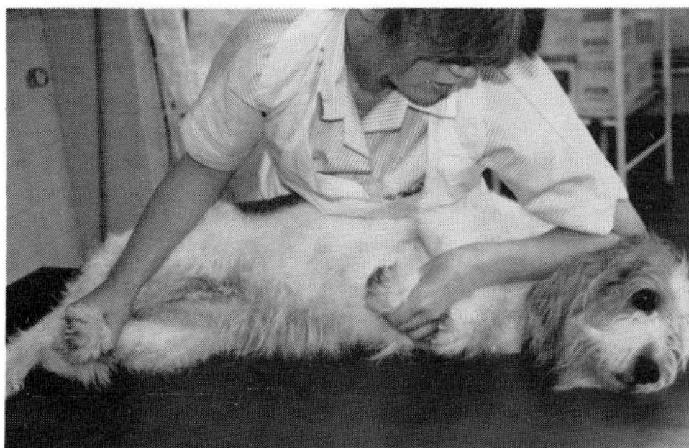

FIG. 11.4 A dog may be placed in a submissive position. It should be grasped firmly, as shown, and slid down gently.

This may even be prevented with a quick slap, but it all depends on the correct interpretation of the signs and the experience of the handler. However, hitting any dog cannot be construed as professional handling. It can easily lead to an embarrassing confrontation with an owner. It is usually better to place the dog in a submissive position (Fig. 11.4).

If the dog is very difficult to handle, prior muzzling by the owner is often worthwhile. Therefore suitable sizes of muzzle should be kept on the veterinary premises. Guidance on how to apply each type of muzzle will usually be needed. Modern plastic and nylon muzzles are effective, light, easily cleaned and sterilized, and are ideal for most purposes (Fig. 11.5). When the dog submits, it should be rewarded with encouraging words and possibly a food reward such as a chocolate drop or other treat.

The dominant position of standing over the dog is augmented by touching. It should be remembered that owners are often bitten by their own aggressive dogs when they are handling them. It is therefore crucial to make sure that a dog is submissive before attempting to stroke or otherwise handle it.

Animals are often apprehensive when attending veterinary premises as they may associate them with discomfort, pain, or at least the experience of being unwell. The induction of discomfort or pain is an extremely dominant act towards a dog. Veterinarians run a high risk of being bitten by inherently dominant animals when they are within the veterinary environment.

Dogs are more confident on their own territory, so it is sound practice to handle dominant animals away from their home. It is sometimes forgotten that even toy or very small dogs can be dominant at

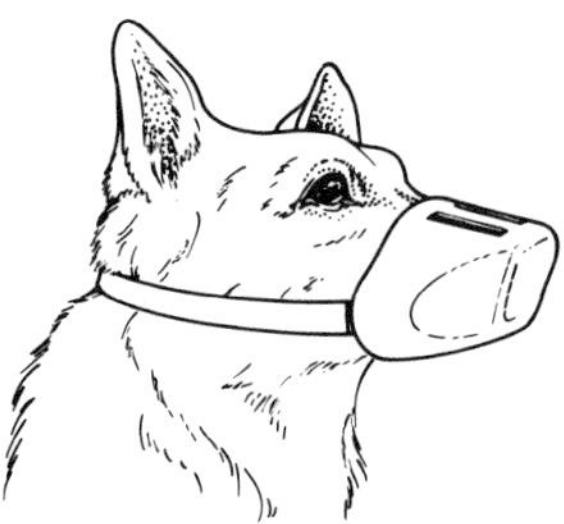

FIG. 11.5 Modern fabric and plastic muzzles are easily applied and are hygienic.

home. It is therefore good practice to ensure that, if hospitalized, they are kept in lower rather than higher cages. They do not then confront the handler at eye level or above, which only adds to the problems of bringing about the animal's submission.

Methods of Approach and Control

General Considerations

Most animals encountered in small animal veterinary practice have received some training and are fairly used to being handled. Exceptions to this are the true strays and semi-feral dogs which are sometimes encountered in and around built-up areas. These present special problems of capture and control and may require measures using chemical restraint, trapping, or noosing with the aid of a dog catcher (grasper).

Strays which have a known history of biting or attack may sometimes be captured after sedation or tranquillization using a dart. On these occasions the local authority, animal welfare society or the police are usually involved.

The majority of dogs can, however, be considered handleable even though they may be apprehensive in a veterinary environment. Body language expressing fear is an extension of the active submission stance. The dog will crouch with the tail held low and will avoid eye contact. Fear is expressed by shaking, shivering and a look of apprehension (Fig. 11.6). The dog will try to hide behind or beneath objects if possible. Care must be taken approaching these animals since they can bite because of fear, often urinating and defaecating at the same time. The approach to any strange dog should be slow, deliberate and gentle. It helps to allow the dog to examine the handler and for the handler to avoid bringing the hands down onto the head from above. It is safer and more effective to bring the hand up to the sternum from below provided the face is not too exposed to the danger of a bite.

Muzzling

Crouching down to the animal's level does much to prevent fear aggression (Fig. 11.7). Fear biters and animals used to a dominant position in the family should be approached with particular care. Prime considerations are the prevention of escape and injury to anyone present, including the owner. In many cases the owner will be able to put a muzzle on their dog unaided, particularly if it is taken out of a potentially stressful environment such as the consulting room, into a quieter area.

A muzzle tends to distract the animal's attention and protects everyone from the sudden bite from a dog which is stressed and may be injured or unwell.

FIG. 11.6 Fear is expressed by avoiding eye contact and holding the tail close to the body.

Fig. 11.7 Crouching down to the animal's level often prevents fear aggression.

It is an infinitely preferable alternative to struggling with an animal. If the owner cannot muzzle the dog but can hold it, simple instructions must be given regarding the method of holding so that a muzzle can be applied. The dog must be held firmly, on either side of the head, just behind the ears. The skin or the hair immediately behind the head must be grasped and, with the palms pressed against the spine, the head is pointed towards the person applying the muzzle. It is usually better to do this at floor level, except with very small dogs. The dog should be approached slowly and deliberately, using familiar words especially the dog's name. The muzzle should be applied and correctly adjusted. If a tape muzzle is applied, it is important that the loops around the muzzle are pulled tightly, with the knot either directly above or below the jaws. Further throws above and below the jaws can then be made with the final throw below the jaws (Fig. 11.8). The ends are then taken behind the ears and tied in a bow in order that a quick release is possible. The muzzle can be prevented from being clawed off if the owner or person holding the dog is instructed to press downwards as well as pointing the head towards the person applying the muzzle. This will prevent the dog from lifting its forefeet to remove the muzzle.

Muzzling is unsuitable for most brachycephalic breeds. Small breeds such as Pekingese, Pugs and Boston Terriers, can usually be restrained effectively with a thick towel placed around the neck behind the ears (see Fig. 11.9). Tape muzzles can be applied to some short-nosed breeds such as Boxers and Bulldogs (Fig. 11.10). Pressure over the bridge of the nose with concomitant respiratory distress can be relieved by passing a piece of bandage under the loop encircling the mouth and then tied behind the head. Tying the ends of this bandage together raises the loop encircling the foreshortened nasal bones. Relieving the pressure on the nasal passages in this way allows the animal to breathe. In practice, this muzzle is particularly difficult to put on and it is more satisfactory to use one of the modern fabric or plastic muzzles, many of which fit most short-nosed dogs.

With any fractious or uncooperative dog, the muzzle must be very securely applied. Tape muzzles are disposable, which is useful when the dog has an infectious condition. Tape muzzles can be made from a piece of bandage. Synthetic materials are stronger. If the owner cannot hold the dog for muzzling, it may be possible to attach two slip

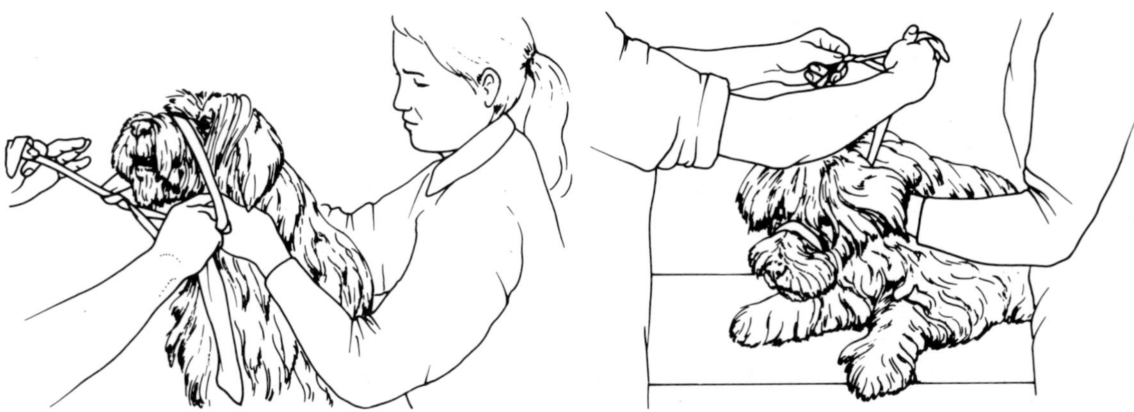

Fig. 11.8 Application of a tape muzzle.

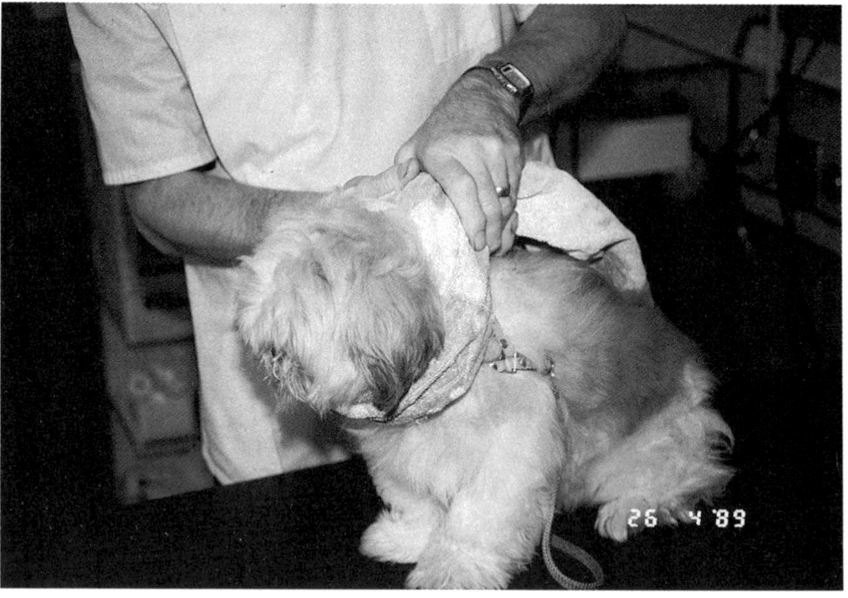

FIG. 11.9 Small brachycephalic breeds can be restrained effectively with a thick towel behind the neck.

leads and, with the dog backed into a corner and approached from the front, a muzzle made of soft rope can be used. Unlike a bandage, this will retain its preformed shape (Fig. 11.11). This should be tied on top of the nose first and then again with a single throw beneath the chin and the ends taken behind the ears, where a quick release bow can be formed. It is then advisable to apply a formed muzzle or another tape fashioned from a bandage. Once the animal is effectively muzzled, further procedures can be attempted with safety.

FIG. 11.10 Pressure on the nasal area must be reduced if a tape muzzle is applied to a member of a brachycephalic breed.

Unruly animals which are loose, even in small enclosed areas, present problems to secure and restrain. Patience is necessary. A quiet, deliberate approach is crucial. It is futile to attempt to lunge at an animal, either with a dog grasper or a looped lead to slip over the head, the likely outcome will be a serious bite. Using the dog's name if it is known and simple commands such as 'sit' and 'stay', which may be recognized by the dog all help. Approach slowly and extend the back of the hand in order that the dog may familiarize itself with body scent. Odour is of greater importance to dogs than humans and fear expressed via our body scent is soon detected by the dog. If the animal shows signs of attack, if may be better to stop and try again later.

Dominant dogs will show signs of attack with ears forward, tail up, hackles raised and a 'four square' stance, often accompanied by growling. Dogs in active submission, adopting a crouching attitude with the ears back and often urinating may

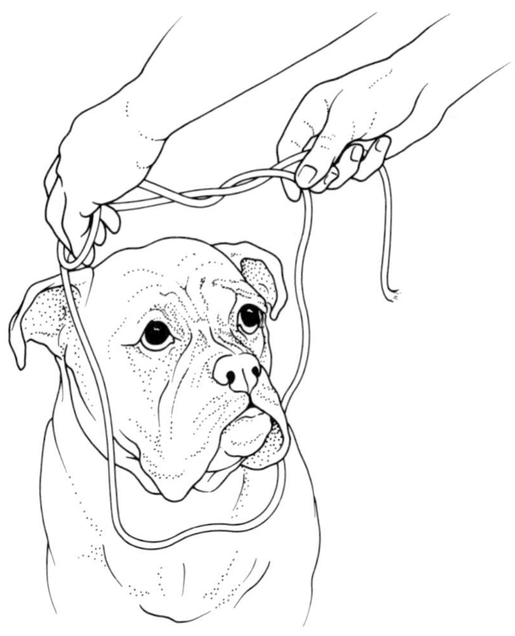

FIG. 11.11 A loop of soft rope will retain its preformed shape.

again attack from fear, particularly if the hackles are raised at the same time.

Dog Graspers

Noosing with the aid of a dog catcher or grasper can be traumatic for both animal and operator and should be used only as a last resort. Dog graspers vary in complexity but essentially consist of a long handle with a loop of rope or other material (Fig. 11.12). One end of the loop is fixed to the end of the handle and the other end runs up the handle so the size of the loop can be controlled. The handle is used to slip the rope over the dog's head from a distance and it is then tightened. Once controlled, a slip lead can then be placed over the dog's head and pulled in the opposite direction, preventing movement until a suitable muzzle can be applied (Fig. 11.13). Alternatively, the dog can sometimes be approached with a pair of thick gloves worn to protect the hands and arms. With patience and gentle deliberate movements, it may be possible to apply a couple of slip leads, or the noose of a dog catcher, over the animal's head. Care must be exercised at all times to prevent the escape of the dog and to prevent personal injury. Speak to the dog quietly throughout the procedure.

Collars and Chokes

Animals which are difficult to control can sometimes be restrained with a choke or check collar attached to a lead. These *must* be applied correctly so that gravity ensures that the untensioned choke slackens (Figs 11.14 and 11.15). A trailing lead attached to the collar when the animal is at exercise often ensures that approach to the dog need not be direct; the lead being picked up first will often result in the dog's submission.

As a general rule a trailing lead should never be left attached to a choke chain lest it become entangled and the dog asphyxiated if left unattended. If the chain is slack it can slip over the dog's head. It is therefore essential to have any lead attached to a correctly adjusted collar. A supply of different sized, clean and sterilized collars, with plenty of holes for adjustment, which can be given to owners to put on their dogs, often avoids many future problems when admitting animals for hospitaliza-

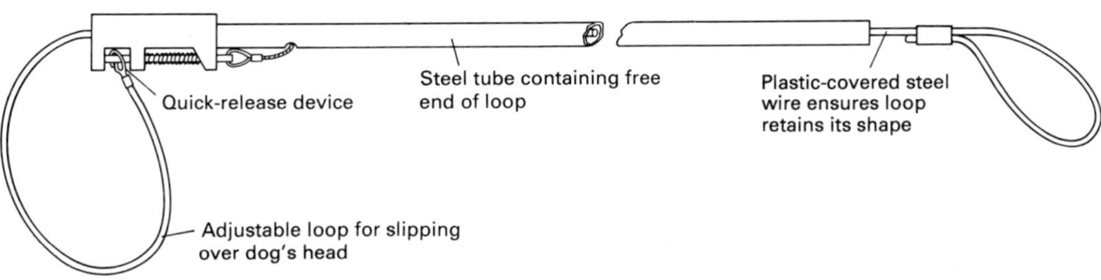

FIG. 11.12 A dog grasper.

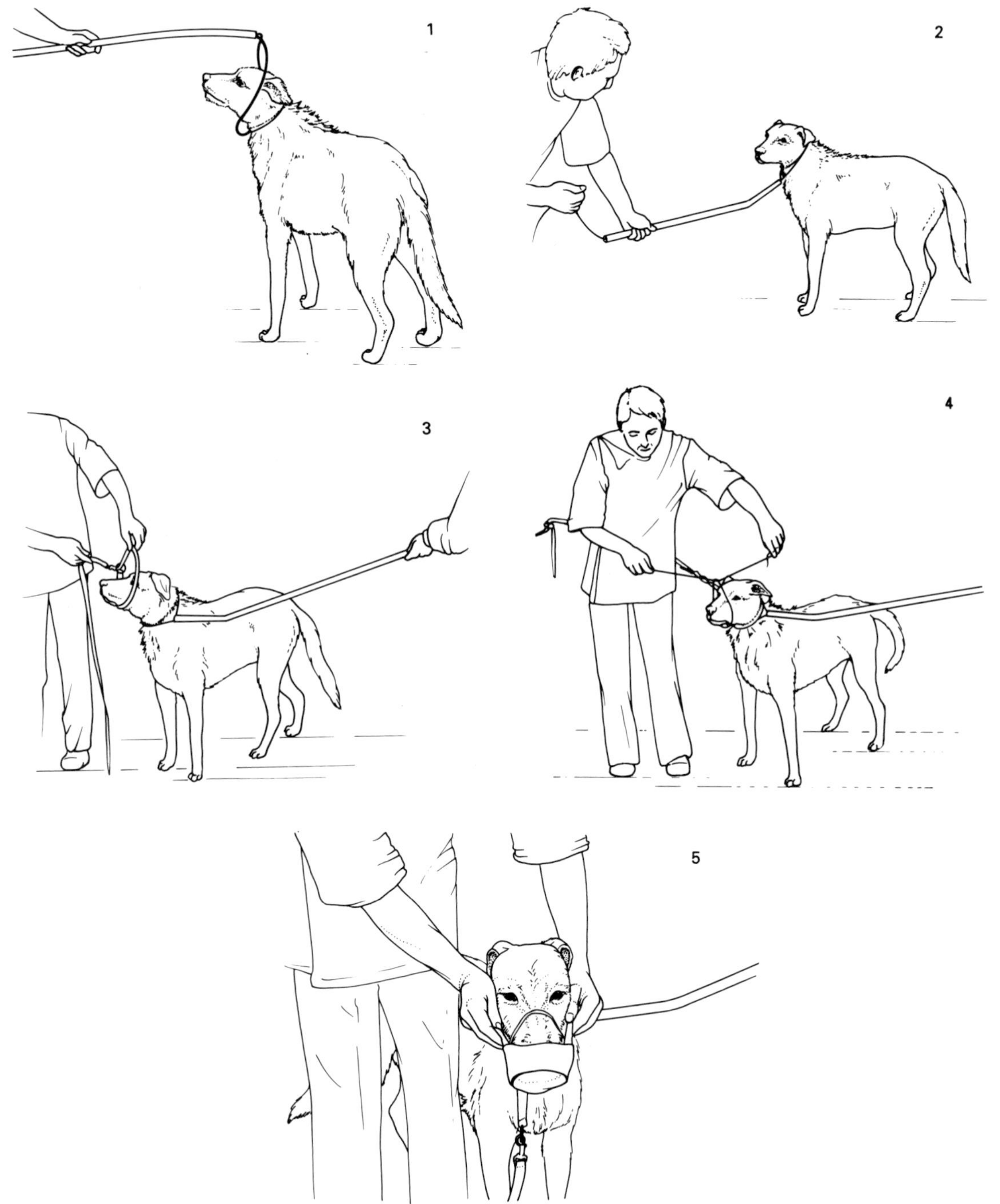

FIG. 11.13 A dog grasper in use.

tion. A hole puncher for leather is a necessity in every practice admitting dogs. Owners always consider that readjustment of the collar will result in it being applied too tightly. Some time will need to be set aside to counsel owners on this subject. Rolled leather collars are not advised as they can stretch with wear and are difficult to adjust properly.

FIG. 11.14 Correct (*right*) and incorrect (*left*) use of a choke chain.

Halter-type Collars

Collars based on the halter principle are often used effectively in dogs. Tension on the attached lead will result in lowering of and turning of the dog's head (Fig. 11.16).

Kennelling

If the dog is to be kennelled in a strange place such as veterinary premises, a lead left attached to a correctly adjusted collar from which the dog cannot escape should be considered, particularly if the animal is uncooperative. If the lead is of leather, there is a possibility that it will be chewed if left attached to the dog for any length of time but leather is more easily cleaned than some woven synthetic types. The latter are also more liable to entangle the dog. A new type of lead made from plastic-covered wire is particularly useful. It can be easily cleaned and left attached with little danger to the patient.

Handling and Restraint for Special Purposes

Lifting and Moving Dogs

Once the dog has been secured (and muzzled if necessary) it often has to be lifted. Dogs up to

FIG. 11.15 Correct (*right*) and incorrect (*left*) use of a check collar.

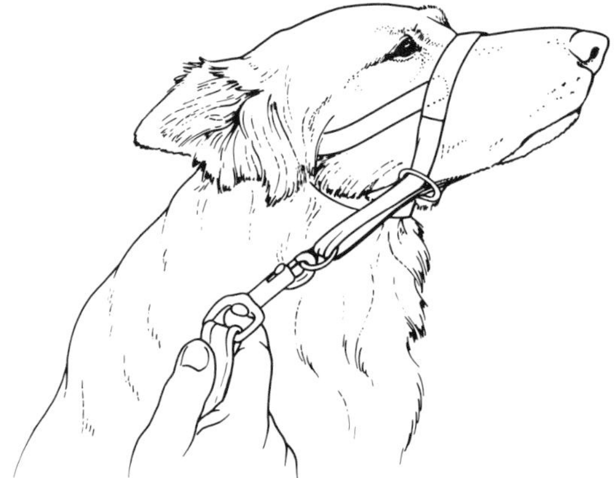

FIG. 11.16 A halter collar. Tension of the attached lead lowers and turns the dog's head.

20 kg can be effectively lifted by one person grasping the animal around the front and rear legs and preventing struggling by pressing the dog's body against the chest (Fig. 11.17). If there is any paraplegia, it is preferable to lift the dog around the chest (the least mobile part of the spinal column) while an assistant places the hindquarters carefully on a table (Fig. 11.18).

Larger dogs may be lifted by one person taking the front end and another the hindquarters (Fig. 11.19). Heavy dogs can sometimes be rolled on to a blanket and then carried by lifting the corners.

FIG. 11.17 Struggling may be prevented by pressing the dog close to the handler's chest.

At least two people are needed for this procedure (Fig. 11.19).

Holding for Examination

There are two common methods of restraint. For most examinations and procedures, such as rectal temperature taking, it is essential that the head is held still. To do this the handler should stand behind or to one side of the dog, which can either be on the floor or on an examination table, depending on size. The collar or hair behind the ears is grasped on each side of the head with the palms pressed against the dorsal part of the neck in order to extend the head (Fig. 11.20). The handler is thus in a good position to adjust the application of gentle to very firm pressure as required.

For examination of the forelimbs and for intravenous injections, the dog should be placed on a table and hugged to the handler who places one arm around the neck and uses the other arm across the dog's back to keep it in a sitting/lying position and also to extend the appropriate limb if required (Fig. 11.21).

Special Restraint for Some Examinations

Restraint on the side

If the dog has to be examined on its side, the fore and hind legs should be grasped by placing the arms across the back and the legs nearest to the handler grasped at the level of the tibia and radius. The legs can then be pulled away from the handler and the body gently lowered against the handler's chest (Fig. 11.22). Once the rib cage is on the arm should be placed across the neck in order to keep the head on the table. The animal is then in the position of submission. This can be stressful for some dogs and a muzzle is advisable and quiet reassurance necessary; although, if the arm is placed firmly across the neck and pressed against the mandible, there is little chance of being bitten.

Restraint on the back

Once on its side, the dog may be rolled on its back and fore and hind legs extended. It may be

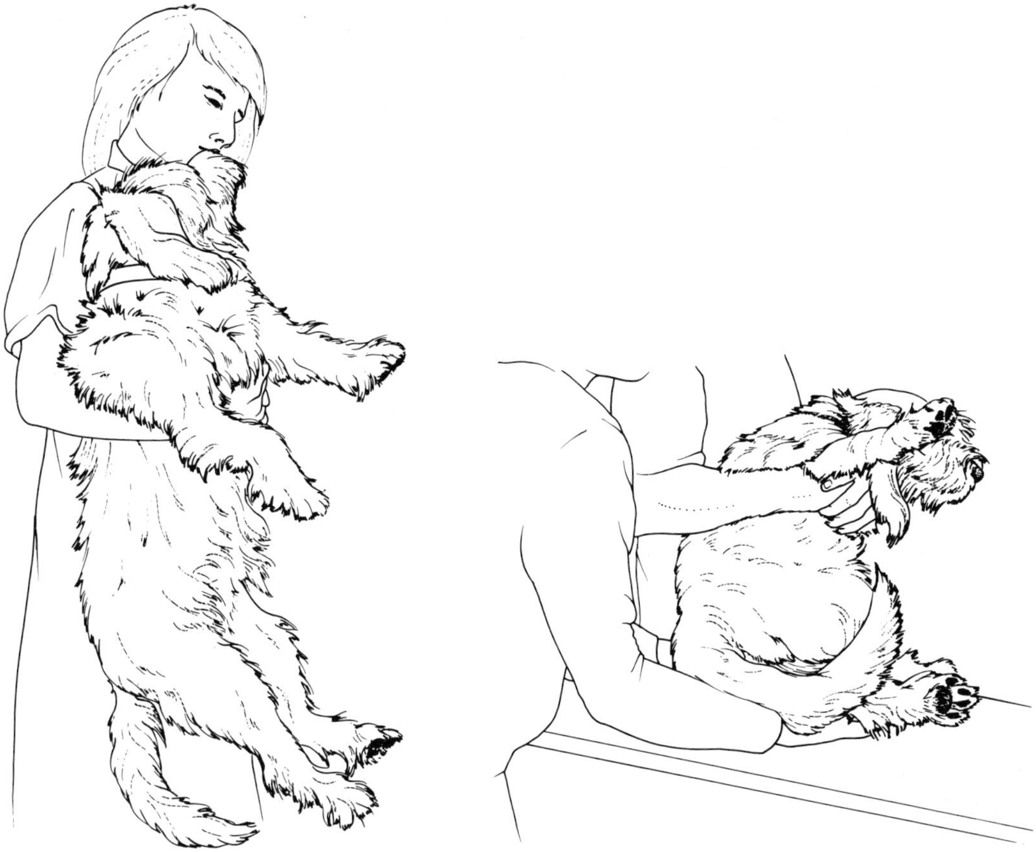

FIG. 11.18 If there is any spinal damage, the dog should be lifted around the chest.

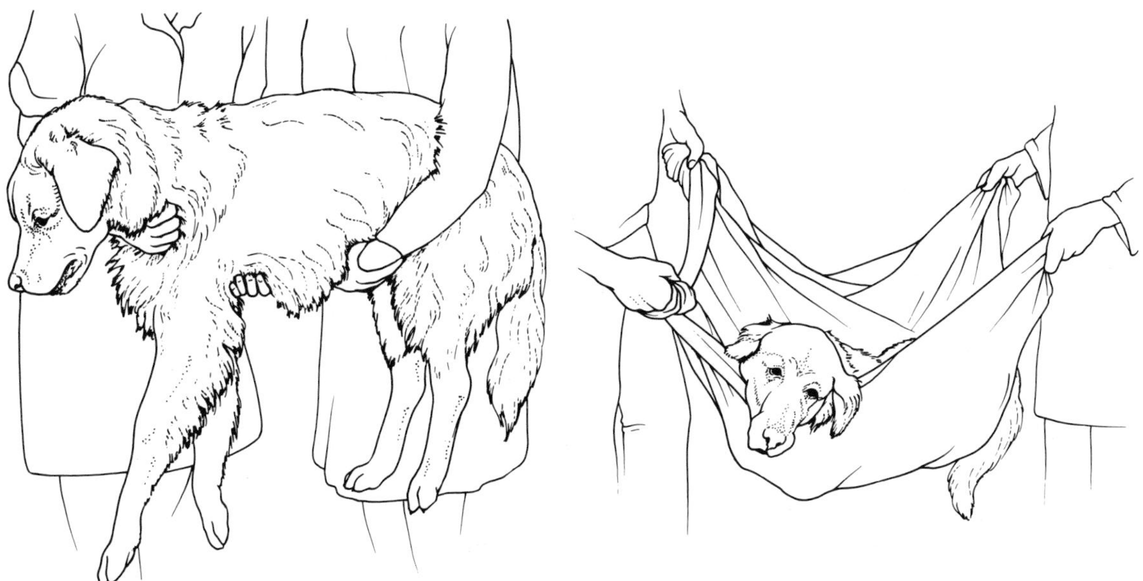

FIG. 11.19 (*left*) Large dogs can be lifted by two people. (*right*) Two or more people can carry a heavy dog in a blanket.

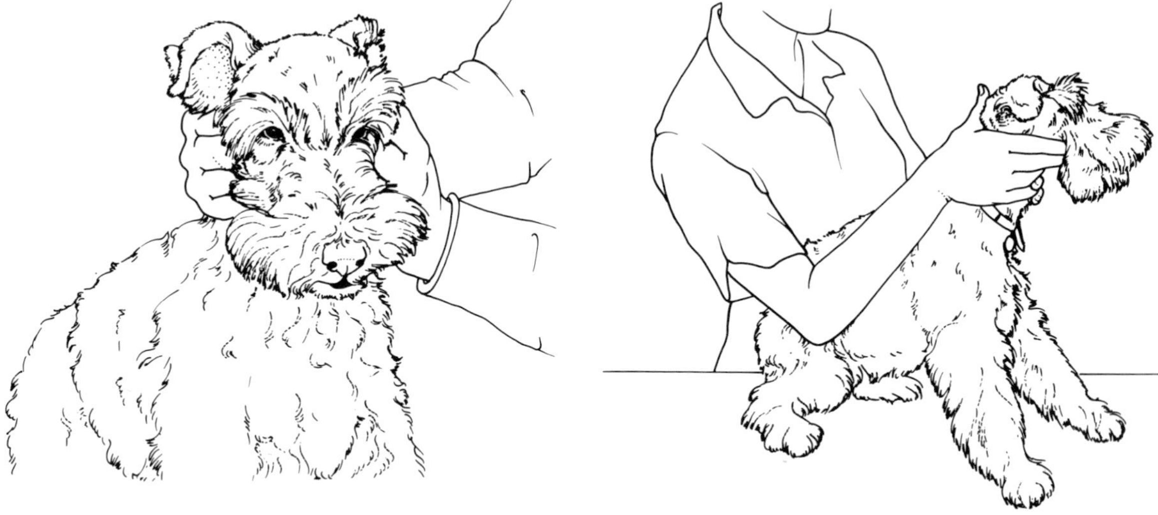

Fig. 11.20 Holding for examination of the head or muzzling.

possible for one person to do this unaided but with larger dogs it is better for two people to take one pair of legs each, the handler holding the forelegs and also grasping the neck between the arms at the same time. Muzzling is advisable as this not only prevents the dog from inflicting injury but focuses attention away from other manipulations being carried out.

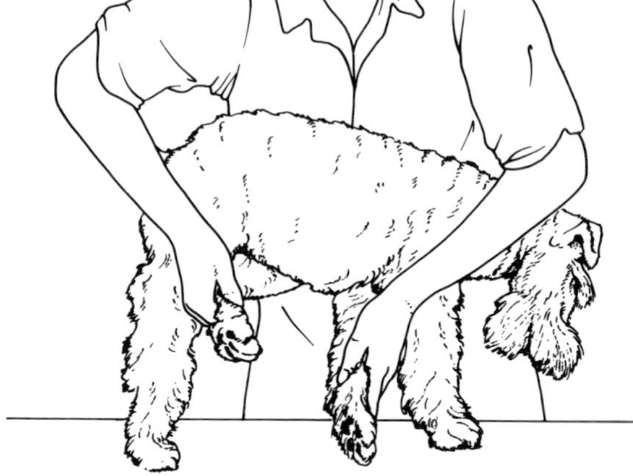

Fig. 11.21 Holding a dog for intravenous injection or examination of the foreleg.

Fig. 11.22 Restraint on the side.

FIG. 11.23 Injection into the hindquarters. This technique is useful for medium or large dogs.

Restraint for injections

Dogs can be held for intramuscular and subcutaneous injections by the first or second method described above, but the second method presents problems if the person is single-handed and the dog is large. Uncooperative animals may be restrained for injection by any of the following methods:

The dog may be straddled and the injection placed in the hindquarters (Fig. 11.23).

If it is impossible to muzzle the dog it may be trapped in a partially closed door (Fig. 11.24). Alternatively, an assistant can restrain the animal's head at the end of a slip lead or dog grasper, while the person injecting the dog pulls the hind leg or the tail, in the opposite direction and makes the injection into the leg muscle, taking care to avoid the sciatic nerve (Fig. 11.25).

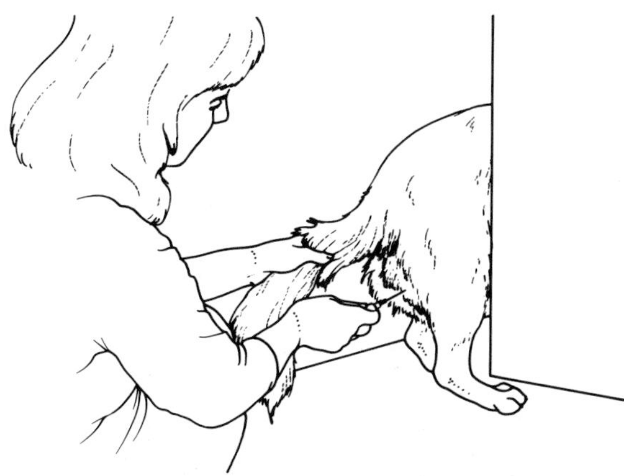

FIG. 11.24 A partially closed door may be used as a means of protection from an unmuzzled vicious dog.

FIG. 11.25 An alternative technique for use with a vicious unmuzzled dog.

Handling and Restraint of Puppies

Neonates cannot regulate their body temperature until the second week of their lives. They are also immunologically incompetent and therefore care should be taken when handling. Scrupulous cleanliness is essential and the ambient temperature should be kept at a minimum of 25 °C. Very small puppies should be handled by grasping with one or both hands entirely surrounding the body. Older puppies should be carried against the chest, remembering that sudden movements are likely. Socialization with people is important from the age of three weeks onwards. Play and handling should be increased each day wherever possible, consistent with the condition.

Handling and Restraint of Geriatric Dogs

Geriatric animals presented for veterinary attention frequently have failing faculties, especially sight and hearing. They are generally less active and adaptable than formerly. They have less ability to withstand stress and therefore handling should be designed to reduce this as much as possible. Discussion with the owner regarding the dog's normal behaviour pattern (irascible, excitable or tranquil) as well as normal feeding and exercise patterns is helpful. A gentle, quiet approach should always be adopted. If the geriatric dog is small and difficult, it can often be handled by grasping through several thicknesses of thick towel or blanket, or by wearing suitable gloves. However, wearing gloves does reduce the ability to feel and it is possible to hurt an animal unintentionally. If elderly dogs have to be hospitalized it should be remembered that they are frequently cachexic and less able to conserve body heat efficiently. Special attention should be paid to providing sufficient bedding and warmth. Adequate bedding is also important to prevent decubital sores, particularly in large and giant breeds. When handling older animals their routine should be disrupted as little as possible. If they have to be hospitalized for protracted periods, they should always be returned to the same cage or kennel and their daily exercise routine should be kept unchanged. This will help to reduce stress associated problems.

Handling and Restraint of Injured Dogs

Owners should be warned that an injured dog is as liable to bite members of its own family as it is strangers. The injured dog should be approached gently and a muzzle applied whenever possible. If the dog is able to walk, it can be secured with a collar and lead or a slip lead. Otherwise it can be carried gently in a manner determined by the location and nature of its injuries. If the hindquarters are injured, it is preferable to hold the dog around the chest, letting the hindquarters hang naturally (Fig. 11.18).

Dogs with spinal injuries should be handled with special care. Pain may be acute on movement and therefore initial muzzling is advisable. Efforts should be made to avoid compression of the spine which can result in more injury to the spinal cord. Carrying the dog by the same method as described above, for those with injuries to the hindquarters, has much to commend it but special care is then necessary not to cause further pain or injury when the animal is placed on any surface, such as the back seat of a car or a table top.

Owners frequently fashion stretchers from pieces of wood or an ironing board in an attempt to plagiarize human first aid methods. Although this is a practical approach, in the conscious animal there is always a fear of the dog moving off the board whilst in transit and injuring itself further. Restraint in the form of elastic luggage retainers is worth considering (Fig. 11.26).

Limb Injuries

The splinting of injured human limbs is a common procedure which has been carried over into veterinary first aid. Splinting in the dog is restricted to the bones and joints below the elbow and the stifle and although these are frequently injured in falls and road traffic accidents, the application of splints in the conscious animal can be traumatic for both animal and attendant. The value in any particular situation should be carefully assessed before any attempt is made at splint application. Muscle spasm following trauma will often 'fix' the limb in the least painful position. Handling should be restricted to the minimum and the dog restrained from unnecessary movement as much as possible, until analgesia/anaesthesia has been carried out.

Handling and Restraint of the Sick Dog

Sick and depressed animals may bite without provocation. They should be handled gently and any procedure completed as quickly and efficiently as possible. Muzzling should be considered. Special care is necessary with animals suffering from elevated temperatures, cardiac or respiratory problems. If the manipulation appears to be causing distress, it is advisable to terminate the procedure.

Transport

If possible, enquire if the dog is used to travelling, since it may then obey simple commands. When transporting a strange dog in a car, it should always be restrained. A lead attached to a suitable

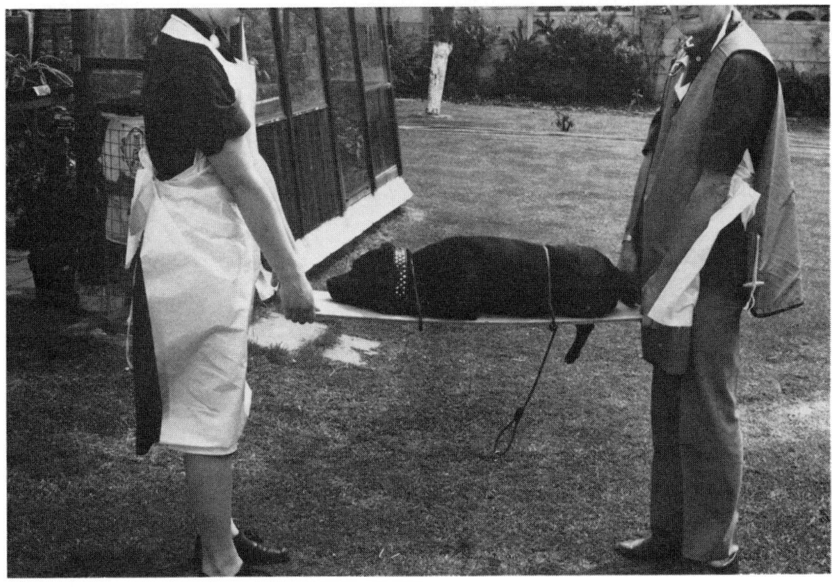

Fig. 11.26 An improvised stretcher; the dog should always be restrained.

projection or shut in the car door is a useful means of restraint. Dogs should not be carried on the back seats of cars as they are liable to be thrown to the floor in the event of a sudden change of speed. It is preferable to transport them on the floor of the vehicle, with the lead sufficiently short to prevent them attempting to climb on the seat (Fig. 11.27). If transported in an estate vehicle, they should be restrained either with the lead shut in the tailgate or attached to a suitable fixing point. Ensure that the lead is sufficiently short that they cannot climb back over the back seat. Alternatively, a barrier in the form of a dog guard is effective (Fig. 11.28).

Sick or injured animals may suffer discomfort if

Fig. 11.27 Shutting a suitably shortened lead in the car door effectively restricts an animal's movement.

Fig. 11.28 A dog guard in use.

the vehicle stops suddenly, or if they try to move around. Such animals need to be carried in a suitable container or travelling box. A simple cardboard box is often sufficient.

Care must be exercised on removal of the dog from the vehicle. If healthy and excited, the dog is quite liable to try to dash out into the road. Shutting the lead in the door offers some advantage. Dogs secured by leads to a fixed point in an estate car frequently attempt to jump from a vehicle when the tailgate is raised and temporarily hang themselves. The lead must be kept short to prevent this. Care in raising the tailgate is necessary. If possible, have someone other than the driver accompany the dog in order that control can be exercised effectively.

Dogs should never be carried in car boots, since they remain unobserved and air supplies are inadequate.

Further Reading

Evans, J. M. and White, K. (1986) *The Doglopaedia*, pp 27–37; 63–70. Henston, Guildford.

Lane, D. R. (1989) Editor *Jones's Animal Nursing*, 5th edition, pp 145–152. Pergamon Press, Oxford.

Lines, S. and Voith, V. (1986) Dominance aggression of dogs towards people, behaviour profile and response to treatment. *Applied Animal Behaviour Science* **16**, (1), 77–83.

Miller, W. C. and Robertson, E. D. S. (1962) *Practical Animal Husbandry*, pp 15–19. Oliver and Boyd, Edinburgh.

O'Farrell, V. (1986) *Manual of Canine Behaviour*, BSAVA Publications, Cheltenham.

Animal behaviour. *Veterinary Clinics of North America* (1982) **12**, (4) 517–533.

The human companion animal bond. *Veterinary Clinics of North America* (1985) **15**, (2).

12

Caged and Wild Birds

JOHN E. COOPER

Introduction

There are nearly 9000 species of bird. These range in size from humming birds to ostriches. They vary in their shape, the extent to which they use their beaks, wings or feet for defence, and in their temperament. As a result, although a number of general rules apply to all birds, there are also specific techniques for handling and restraint which may need to be mastered. In this chapter the term 'handling' is used in its dictionary sense of 'touching, feeling or taking in the hand' and 'restraint' to mean 'keep in check or under control'. The latter term is therefore taken to include techniques such as the wrapping of birds in towels or specially designed jackets in order to carry out a clinical examination, to take samples, or to administer treatment.

Some of the main groups of birds and the ones on which emphasis is laid in this chapter, are listed in Table 12.1.

TABLE 12.1 *Some examples of birds*

Order	Groups covered	Examples
Passeriformes	'Perching birds' such as thrushes, starlings, finches	Greater hill mynah (*Gracula religiosa*) Canary (*Serinus canaria*)
Psittaciformes	'Psittacines' — parrots and their allies	African grey parrot (*Psittacus erithacus*) Budgerigar (*Melopsittacus undulatus*) Cockatiel (*Nymphicus hollandicus*)
Falconiformes	'Diurnal' or 'falconiform' birds of prey such as hawks, eagles and falcons	Kestrel (*Falco tinnunculus*) Sparrow-hawk (*Accipiter nisus*) Golden eagle (*Aquila chrysaetos*)
Strigiformes	'Nocturnal' or 'strigiform' birds of prey — owls	Tawny owl (*Strix aluco*) Barn owl (*Tyto alba*)
Columbiformes	Pigeons and doves	Domestic pigeon (*Columba livia*) Wood pigeon (*Columba palumbus*)
Anseriformes	Ducks, geese and swans	Mallard (*Anas platyrhynchos*) Mute swan (*Cygnus olor*)
Galliformes	Gamebirds	Japanese quail (*Coturnix coturnix*) Peafowl (*Pavo cristatus*)

General Points

The majority of birds, even if gregarious in the wild, will resist being handled, although some tolerate limited physical contact such as stroking. Hand-reared birds, especially if 'imprinted', will prove most amenable to touch and restraint.

Diurnal birds can often be handled and restrained more easily in subdued light. Nocturnal species, on the other hand, may prove more tractable in a bright light.

Maximum dexterity is achieved if birds are handled without gloves, but under certain circumstances these and other aids (Table 12.2) may be needed and should be both chosen and used with care. During handling the fingers should be separated so as to minimize pressure on the bird's body.

As a general rule the aim when handling birds is to restrain the wings around the body. There are, however, exceptions and it should be noted that in Britain The Welfare of Poultry (Transport) (Amendment) Order 1989 allows domestic geese to be lifted and carried by the base of both wings, and ducks by the neck.

The plumage of birds is important — for insulation, for optimum flight and to enhance appearance. Great care must be taken not to damage feathers when handling birds, especially those species which are used in sport (e.g. pigeons, hawks) or exhibited (e.g. show budgerigars, canaries). The use of a suitable net (of correct size and handle length and with a padded rim) is helpful in this respect.

Although many species can be restrained adequately by physical means, chemical agents are often useful — and in the case of large species, sometimes essential — if clinical examination is to be performed.

Birds can easily escape or damage themselves. The risks will be minimized if careful preparation is made (see later).

Welfare is of great importance and must be borne in mind whenever a bird is handled or restrained. Assessment of welfare is not easy:

TABLE 12.2 *Equipment which may facilitate handling and restraint of birds*

Equipment	Purpose	Comments
Gloves	Reduction of damage to handler	Avoid unless essential. Use thin gloves wherever possible: even rubber gloves will minimize wounds. Elbow length gloves can be useful for large aggressive birds
Towel/cloth	To wrap around bird in order to facilitate handling and permit restraint for examination/sampling/treatment	An invaluable aid. Various thicknesses (one or more folds) can be used for different purposes
Clothbag/sack/stocking/pillow case	To place bird in so as to minimize struggling and to facilitate weighing and other procedures	Care must be taken not to asphyxiate or damage the bird
Cardboard tubing	As above	Frequently used by field biologists in North America. The bird appears quieter and less easily stressed
Hood	To cover head of (diurnal) bird in order to reduce stress and trauma	A standard method of quietening and restraining falconers' birds: can be used to advantage in many other species. A well fitted hood is preferable to a loose cloth bag
Harnesses and other devices	To restrain bird so as to minimize struggling and facilitate procedures	Many designs available (see references), including the 'Guba' used for falconers' birds (Figs 12.9, 12.10)
Elastic bands and sticky tape	To seal beak and to protect the handler	Remember 1) that the bird can still stab and 2) to remove band or tape before release

recent studies on ducks have attempted to compare the effects of different methods of restraint on blood enzymes (Bollinger *et al.*, 1989). Such an approach may help in assessing techniques in the future.

Preparation

Before a bird is handled prior to clinical examination, it should be observed in its cage or enclosure. This will permit subtle clinical signs to be detected which may assist diagnosis. In addition, such observation may provide information as to precautions which should be taken during handling and restraint. For example, a bird with an increased respiratory rate may need to be particularly carefully handled and not restrained for too long if dyspnoea is to be avoided. Likewise, a bird may have a visible wound or lesion, in which case a method of handling may need to be chosen which will minimize trauma and possible bleeding. Observation, coupled with questioning of the owner, will also help in elucidating whether a bird is capable of full flight (and thus be a potential escape risk) or, perhaps, pinioned or feather-clipped.

Techniques for handling will differ, depending upon whether the bird is in a cage or aviary. Before removing a bird from a cage (or other small container) it is essential to check that all windows and doors are closed, extractor fans are turned off, possible exits (including chimneys and fireplaces) are blocked, glass panes against which the bird may stun itself are protected and no other animals,

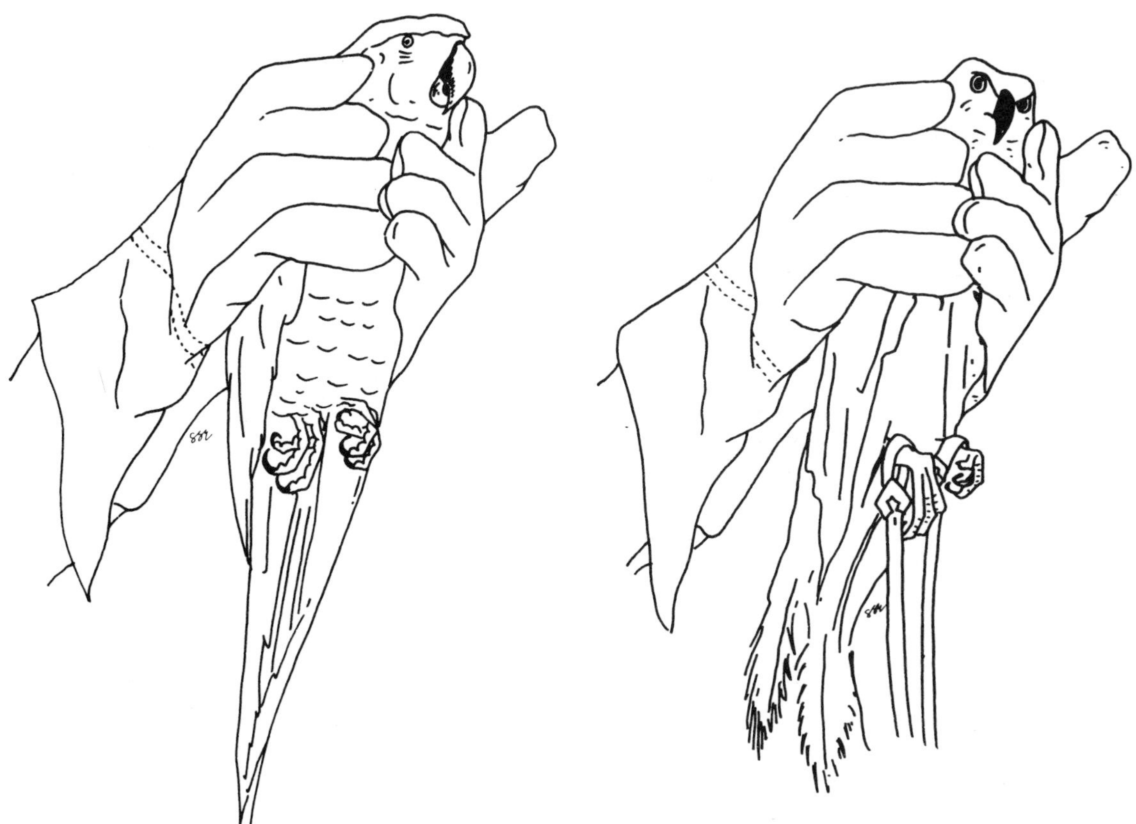

Fig. 12.1 (*left*) A small macaw is handled with the aid of gloves. In such species, the beak is more dangerous than the feet: the head should either be restrained with the fingers (as illustrated) or grasped from behind by a second handler. (*right*) Gloves are also used to handle a hawk. Note the jesses hanging from the feet; these can be grasped to increase control over the bird and to minimize damage to the claws (talons).

FIG. 12.2 A goose is restrained using two hands. Note how the bird's body is supported with one arm.

FIG. 12.3 A quail is handled in order to examine the plumage and dorsal surface of the wing. Note how one hand restrains the body while the other grasps the carpal joint and gently extends the wing.

e.g. dogs or cats, are loose in the room. Birds which are caught in an aviary should be transferred in a box, cage, net or bag to the examination room and not be carried in the hand. The only exception to this are birds which have been pinioned (and are therefore unable to fly away) or species/individuals which jump or flap excessively and can damage themselves in a box or cage.

The sequence of events that should be followed when handling a caged bird is given in Fig. 12.4. Equipment which may facilitate handling and restraint is also listed in Table 12.2. Prolonged chasing can prove deleterious and every effort must be made to minimize the capture time, especially when dealing with sick birds: a net is particularly valuable in this respect.

Physical Restraint

Handling

As emphasized earlier, the main aim when handling birds is to immobilize the wings (Fig. 12.5). This will minimize flapping and as a result there will be less danger of trauma and stress, both to the bird and the handler. Small species can be grasped and then held in one or both hands, but larger birds may need to be restrained using a towel.

An important point when capturing a bird in an aviary, or a cage containing more than one species, is to take careful note of the bird to be caught. This is because a sick bird may, during a chase, temporarily appear fit and well and as a result prove difficult to identify.

Full use should be made of subdued lighting, as mentioned earlier. Small birds can often be lifted off their perches, with the minimum of disturbance, if the procedure is carried out in a darkroom using only a small light source. Alternatively — in an aviary, for example — handling may prove easiest if carried out at night, although not if there is a full moon or light from a nearby building, when birds may panic and damage themselves. Care must always be taken to avoid disturbing other birds in the same or an adjacent aviary, as these may be disturbed and either injure themselves or roost in an exposed site. Certain owls, nightjars and other nocturnal birds will often appear dazed in bright light: they are best handled during the day or with the aid of a strong lamp.

TABLE 12.3 *Methods of handling and restraint*

Group	Main points	Additional points
Small passerines	Grasp in hand or net. Hold in one hand with 2nd and 3rd fingers around head and thumb and 4th and 5th finger around body (Fig. 12.6); release fingers in order to examine wings or take samples (Figs 12.3 and 12.12)	May stab or bite with beak: thin gloves will help to minimize effect. Use elastic band or sticky tape to seal beak
Large passerines	Hold with two hands, round wings (Fig. 12.5). Place on a towel on flat surface to examine wings or to take samples	As above. Light (gardening) gloves may facilitate handling
Small psittacines	As for small passerines	As for small passerines but less inclined to stab. Usually not practicable to seal beak: best to restrain head with other hand or to cover with a cloth or small bag
Large psittacines	As for large passerines. Examination and sampling may necessitate chemical restraint	As for small psittacines: head will need to be restrained or covered by a second person
Small and medium birds of prey (falconiform and strigiform)	As above (large passerines)	The claws (talons) usually present more of a hazard than the beak. Light gloves will minimize effect. Falconers' birds can be handled easily if hooded. Jesses and leashes can be used to advantage to facilitate examination and sampling (Fig. 12.1). Avoid damaging plumage
Large birds of prey	As for small and medium birds of prey. Can use cloth to grasp round wings. Alternatively catch while bird perching by seizing legs and quickly turning bird upside down: the wings will usually be extended but can be readily folded in to the body (Fig. 12.1)	The feet can be hazardous and it may prove very difficult to loosen the bird's grip without levering out the talons one by one. Use heavy (reinforced) gloves and, where appropriate, falconers' equipment
Pigeons and doves	As for small/large passerines. Pigeon fanciers prefer to hold birds with one hand, around the base of tail	Rarely bite or scratch. Inclined to defaecate during handling. Feathers readily lost — try to minimize this and damage to plumage in racing bird
Waterfowl	As for large passerines	May bite: some geese have sharp claws and powerful legs and can inflict severe scratches. Swans and geese may flap wings and prove difficult to restrain (Fig. 12.2)
Gamebirds	As for large passerines	May bite, stab with spurs or scratch with claws. Some species e.g. quail, inclined to leap into the air and may concuss themselves (Fig. 12.3)
Waders Herons Storks Cranes	As above, depending on size. Grasp neck of herons, storks and cranes first in order to restrain head (beak) (Fig. 12.7)	May stab with beak: protect eyes and exposed skin. Handle with care as long legs prone to damage, including fractures. Storks and cranes have strong legs and will kick
Gulls Terns Shearwaters Petrels	As above, depending on size	Gulls very likely to stab with beak: always use an elastic band. All this group inclined to vomit during handling: fulmars may regurgitate oil

N.B. Some birds will remain stationary for a short time, if placed on their backs. However, such 'tonic immobility' may be unnecessarily stressful to the bird as well as creating a risk of escape.

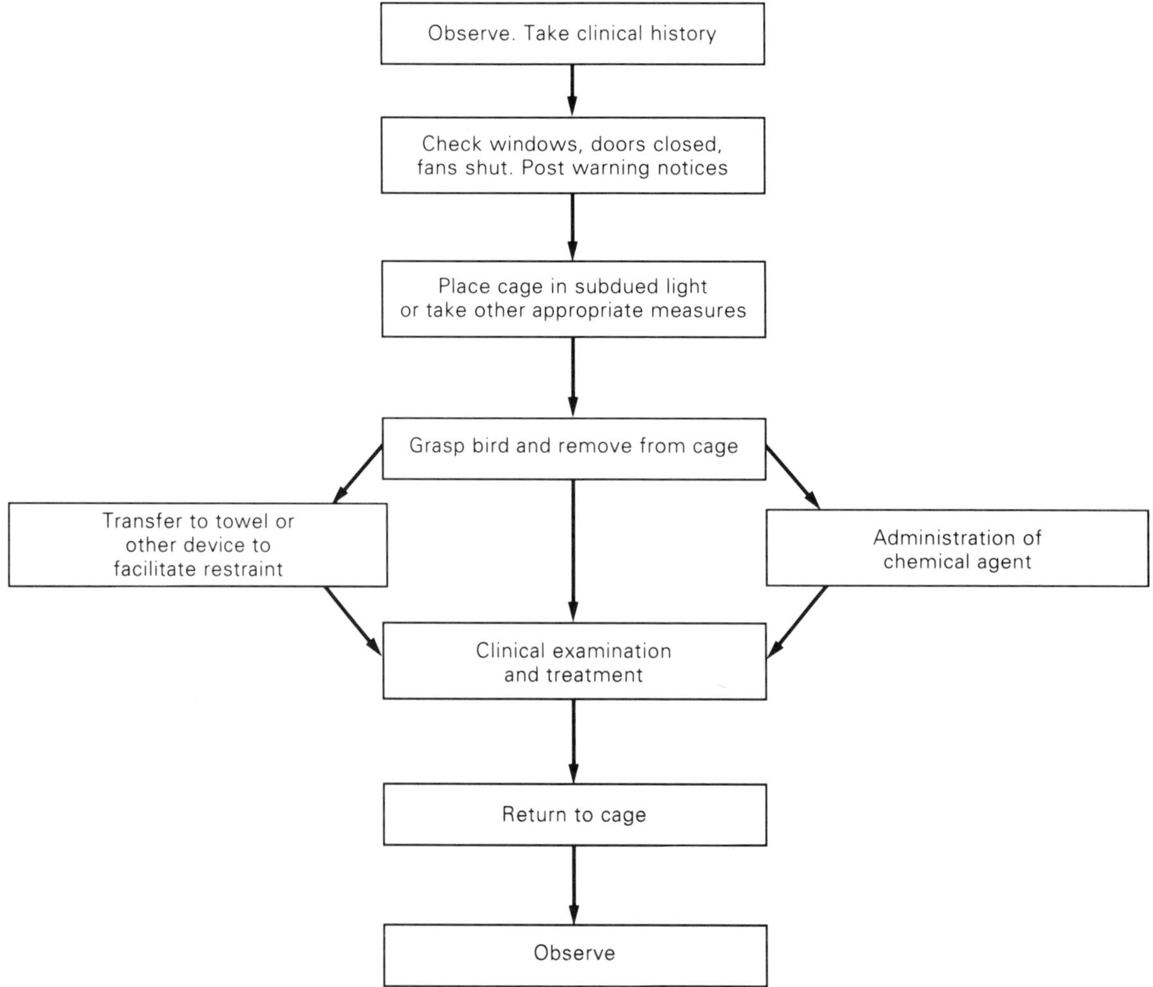

FIG. 12.4 The sequence of events to be followed when handling caged birds.

Restraint

Various techniques and items of equipment can be used to restrain birds for examination or in order to take samples (Table 12.2).

Handling and Restraint of Specific Types of Birds

Recommended techniques are given in Table 12.3. Reference should be made to this and to the appropriate figures.

Chemical Restraint

Although this subject is outside the remit of this chapter, chemical restraint is so relevant to the handling of birds that brief mention must be made of it. Injectable agents can be used, e.g. ketamine hydrochloride, or inhalation agents, e.g. methoxyfluorane — for example, delivered by means of an anaesthetic chamber (Cooper, 1989; Applebee and Cooper, 1989). The anaesthetic chamber also serves as a useful box for observation and examination.

Accidents and Emergencies Associated with Handling and Restraint of Birds

Accidents will be minimized if the points listed at the beginning of this chapter are followed. Very occasionally a bird dies in the hand for no apparent

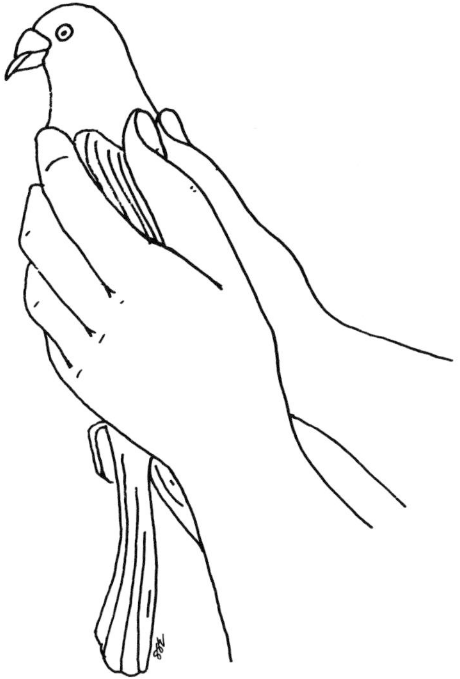

FIG. 12.5 A pigeon held in the hand. Note how the two hands encompass the wings and prevent the bird from flapping.

FIG. 12.7 A crowned crane is held under the arm. One hand grasps the top of the neck, to discourage pecking and uncontrolled movement of the head, while the other holds the legs together.

reason but usually there is a clear cause, such as undue pressure on the body (often because the bird is struggling) or underlying disease. Sometimes a leg (less frequently a wing) is broken. Prolonged restraint of large birds may result in limb paralysis or impaired function. Capture myopathy has been reported in the larger wild birds, such as flamingoes, following a prolonged chase, poor handling or transportation.

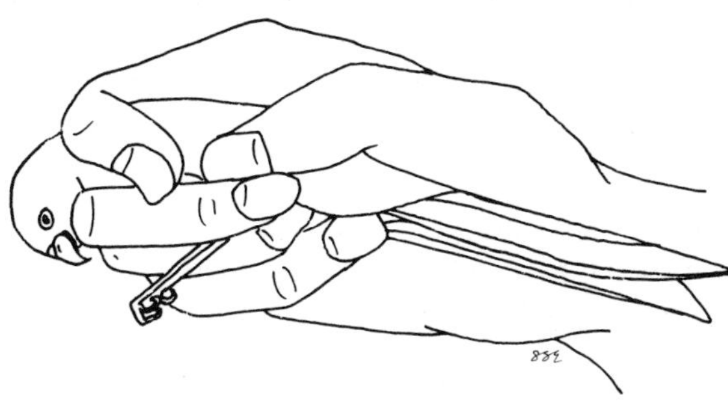

FIG. 12.6 A budgerigar in the hand. Note how the fingers form a 'net' around the bird: undue pressure must not be applied.

FIG. 12.8 Removal of a bird from a cage can be stressful for both the handler and the bird. The situation is compounded by certain designs of cage, such as this old-fashioned round one. Under such circumstances subdued lighting will help.

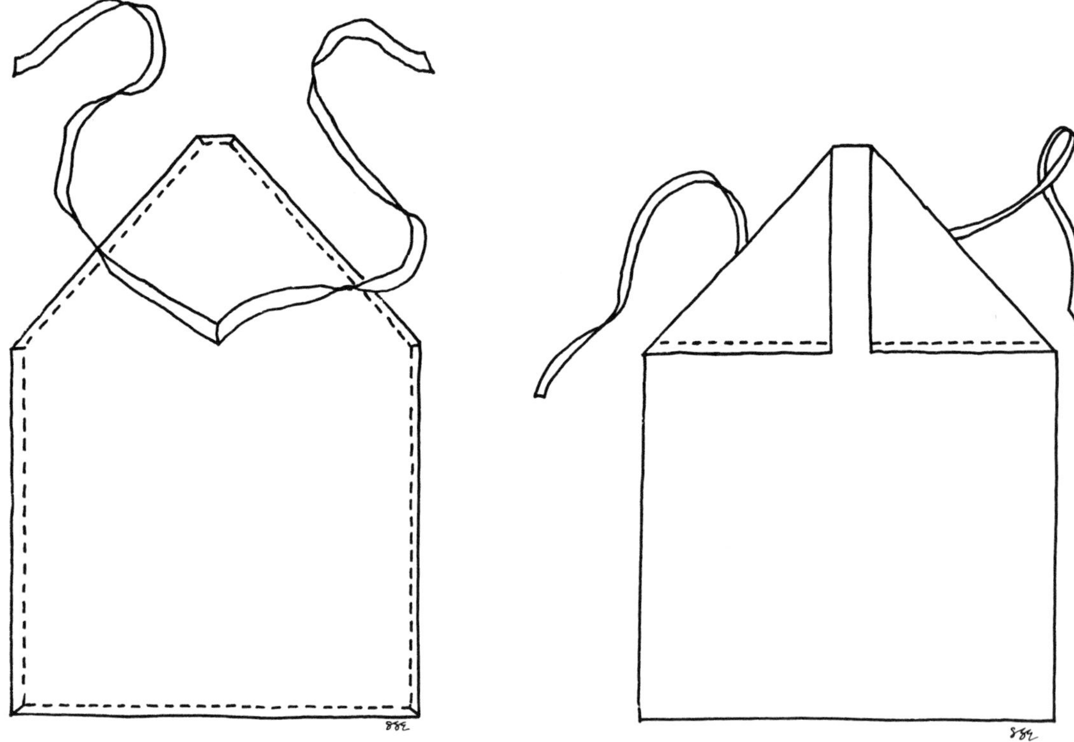

FIG. 12.9 Various devices can be used to restrain a bird and to facilitate the taking of samples. This is the 'Guba', a simple harness used by the Arabs. The bird's wings are placed in the two folds shown in the picture on the right (Cooper and Al-Timimi, 1986).

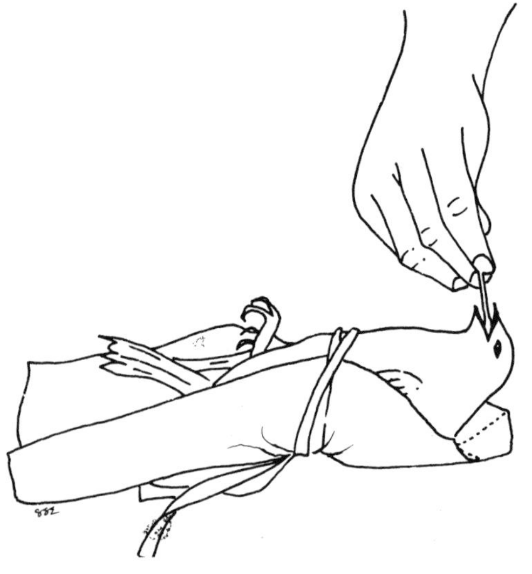

FIG. 12.10 The 'Guba' in use. Note how the bird can be examined and sampled by one person.

Accidents to humans are also outside the main remit of this chapter but brief mention should be made of the danger of bites, scratches and damage to the eyes. Appropriate precautions should be taken and, if accidents do occur, prompt attention must be paid to treatment. In addition, it should be remembered that larger birds may cause injuries with their wings (e.g. swans) or by kicking with their legs (ratites, e.g. ostriches, emus). Precautions which may be taken with the latter are outlined by Meij and Lumeij (1989).

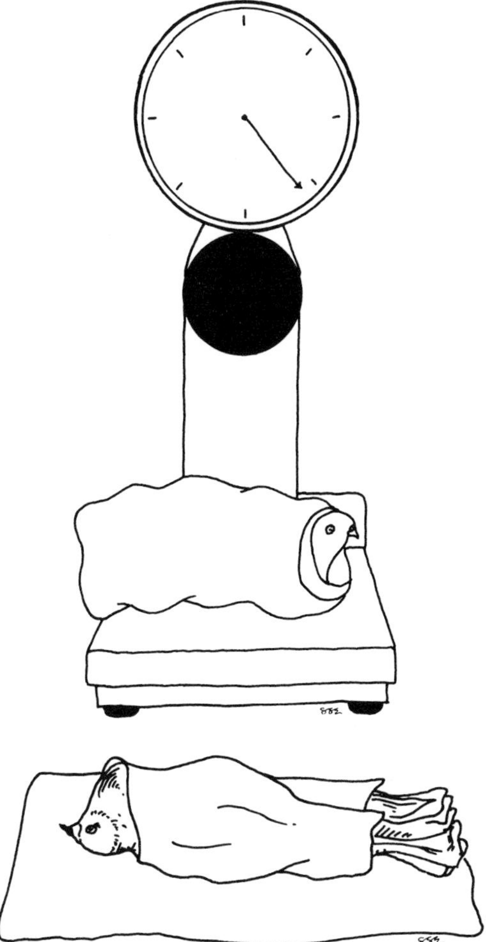

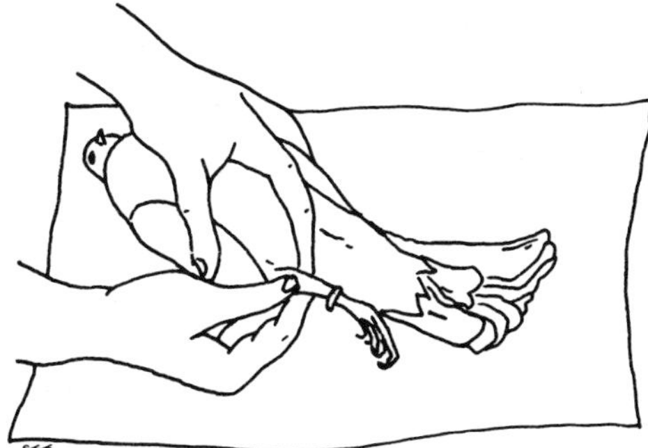

FIG. 12.13 The ring on the leg of a bird is examined. Note how one person can restrain the bird in this position so long as the fingers are parted in order to cover the body and wings.

FIG. 12.11 A bird is weighed. If firmly wrapped in a towel (*above*), many birds will cease flapping. Alternatively (*below*), the bird can be placed in a close-fitting stocking for greater security.

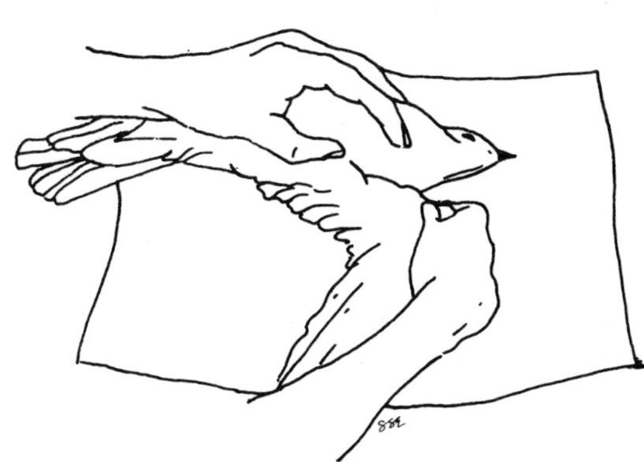

FIG. 12.12 A bird is placed on its back in order to reveal the brachial (basilic) vein, just distal to the elbow joint.

References

Applebee, K. and Cooper, J. E. (1989) An anaesthetic or euthanasia chamber for small animals. *Animal Technology* **40**, 39–43.

Bollinger, T., Wobeser, G., Clark, R. G., Nieman, D. J. and Smith, J. R. (1989) Concentration of creatine kinase and aspartate aminotransferase in the blood of wild mallards following capture by three methods for banding. *Journal of Wildlife Diseases* **25**, 225–231.

Cooper, J. E. (1989) Anaesthesia of exotic species. In *Manual of Anaesthesia*, Editor Hilbery, A.D.R. BSAVA, Cheltenham.

Cooper, J. E. and Al-Timimi, F. (1986). A simple restraining device for birds. *Avian/Exotic Practice* **3**, 5–7.

Meij, B. P. and Lumeij, J. T. (1989) Basic avian handling techniques. In *Proceedings of 2nd European Symposium on Avian Medicine and Surgery*, Utrecht, March 8–11, 1989.

Further Reading

Bloom, P. H. (1987) Capturing and handling raptors. In *Raptor Management Techniques Manual*, edited by B. A. Giron Pendleton, B. A. Millsap, K. W. Cline and D. M. Bird, National Wildlife Federation Scientific and Technical Series No. 10. Washington DC.

Coles, B. H. (1985) *Avian Medicine and Surgery*. Blackwell, Oxford.

Cooper, J. E. (1978) *Veterinary Aspects of Captive Birds of Prey*. Standfast Press, Gloucester.

Cooper, J. E. (1983) A practical approach to cagebirds. *In Practice* **5**, 29–33.

Cooper, J. E. (1984) A veterinary approach to pigeons. *Journal of Small Animal Practice* **25**, 505–516.

Cooper, J. E. and Eley, J. T. (1979) Editors *First Aid and Care of Wild Birds*. David and Charles, Newton Abbot.

Ensley, P. (1979) Caged bird medicine and husbandry. *Veterinary Clinics of North America: Small Animal Practice* **9**, 499–525.

Evans, M. and Kear, J. (1972) A jacket for holding large birds for banding. *Journal of Wildlife Management* **36**, 1265–1267.

Fuller, M. R. (1975) A technique for holding and handling raptors. *Journal of Wildlife Management* **39**, 824–825

13

Reptiles

MARTIN LAWTON

Introduction

Reptiles are members of one of the oldest classes of the animal kingdom, and can be traced further back in time than the origins of mammals or birds. There are approximately 6500 species of reptiles and they are divided into four orders.

The order Rhinocephalia consists of only one species, the *Tuatara*. This is only found in islands off the coast of New Zealand and is most unlikely to be encountered elsewhere due to export bans.

The order Crocodilia consists of 22 species and includes the crocodiles and alligators. All these creatures are covered by the Dangerous Wild Animals Act 1967 in the United Kingdom.

When it comes to pets, the order Chelonia is probably the most popular of the reptile group although there are only 244 species. These include the turtles, terrapins and tortoises. They are set aside from the other reptiles in that they have an external shell. They may be aquatic, terrestrial, or a combination of both.

Squamata is the largest order with 6280 species. This order includes the lizards, worm lizards and snakes.

All reptiles are poikilotherms (ectotherms) and are therefore reliant on the external temperature for maintaining their metabolism. All reptiles have a preferred body temperature (PBT). This is the temperature range at which they thrive best. The PBT varies with the seasons, and may even vary at different times of the day.

When a reptile is being kept at its PBT, its metabolism and enzyme activity are optimal. It is also at its most active, able to reproduce best, and is most resilient to disease, due to immunoglobulin production. The majority of illnesses that are seen in reptiles are associated with the reptile being kept below its PBT.

When examining or handling reptiles, some use may be made of the fact that they are affected by their external temperature. If a normal clinical examination is being carried out by a veterinarian, there is a considerable advantage in the animal being examined at its PBT, as then it will be showing its normal state of activity and it will be easier to see if it is subdued or ill. However, sometimes it is an advantage to handle reptiles below their optimum temperature as they will be slowed down to a certain extent. Mild cooling for handling may be used, but this must be done with extreme caution as it can also damage the reptile.

All reptiles are able to shed their skin. Care must be taken when handling a reptile about to shed, so that the underlying developing skin is not damaged.

The approach towards handling reptiles will be dealt with under four headings: Chelonia, snakes, lizards, and Crocodilia. An important point to remember is that everything is slow in reptiles, and therefore patience may be needed in order to handle and examine members of the reptilian class safely and correctly.

Chelonia

All Chelonia have an external part to the skeleton, known as the shell. This is divided into two parts, the upper part being the carapace and the lower part the plastron. There is species variation in the size of both and in some they may be very reduced. Some species, such as the American box tortoise (*Terrapene* spp) and African hingeback (*Kinixys* spp) are able totally to close up in their shell, by utilizing a hinged plastron or carapace, respectively. On handling these species, care must be taken not to get a finger trapped. Some Chelonia have naturally soft shells, e.g. the *Trionyx* spp. These must be handled with care, partly because of their ferocity, but also to prevent damage to their shell.

FIG. 13.1 A *Trionyx* extending its neck threateningly.

The shell is a means of protection and, to a certain extent, this can prove both an advantage and disadvantage in the handling of these creatures.

Members of the Chelonia have no teeth, but a hard keratin beak, which can inflict a painful bite, especially where the beak is serrated. Some species, especially terrapins and turtles, have long claws and care must be taken on handling not to get scratched. It must also be remembered that all Chelonia have very powerful neck muscles and some, e.g. snake necked turtles (Chelidae), or snapping turtles (*Chelydra*) can extend their necks quite far out (Fig. 13.1), in order to bite the person trying to handle them.

Approach and Handling

As pets, tortoises are used to being handled, whereas most turtles and terrapins are not. This should be remembered before approaching and handling. Tortoises, if frightened during handling, are more likely to withdraw into their shell, and generally offer little aggression. Terrapins and turtles, however, may be more aggressive, and may try to bite rather than withdraw into their shell. Therefore more care has to be taken with these species.

When dealing with tortoises, it is generally advisable to pick them up around the middle of the shell, supporting the whole of the body. With terrapins, it is advisable to pick them up round the back end of the carapace, with the fingers in the inguinal area (Figs 13.1, 13.2). This is to avoid the claws and beak, especially of those which have long necks.

If one is dealing with a terrapin in a tank, a net can be used for a small animal. A tea-towel or towel can be inserted into the water to push a larger creature into the corner of the tank before taking hold of it round the back of the carapace.

Sometimes, when dealing with the aggressive *Trionyx* spp, *Chelydra* or Chelidae, it may be advisable to place a tea-towel or blanket in front of the head, to absorb a bite. These species, once they bite on to something have a tendency not to let go but to bite harder. It is therefore better to offer something inanimate rather than risk a finger.

Common Manipulations of Chelonia

Care and patience are required when examining or manipulating Chelonia. Tortoises that clamp up in their shell may be extruded from their shell by

FIG. 13.2 Holding a red-eared terrapin from the rear/middle of the carapace.

applying gentle pressure and fatiguing muscles. Jerking against them, may result in damage.

Examination of the head of a tortoise that has clamped up is a two-person task. First, one foot should be moved towards the mid-line of the front of the carapace (Fig. 13.3), so that the fingers may be put around the leg and the leg gently pulled out, retaining it against the side of the shell and then repeating the process on the other leg. Once the legs are out and held, an attempt may be made to withdraw the head by placing the finger and thumb on each side of the head, and inverting the tortoise, so that the head comes downwards. With patience the head can be grasped behind the occiput, and then, by applying gentle pressure, the head can be extruded from within the shell. In the case of difficulty, whelping forceps can be employed. These should be used gently, or crushing damage may result. The jaws of the forceps should be placed below and above the head. The head is grasped and gentle traction applied, until it is out and held. In order to look into the mouth, the fingers are placed on either side of the mandible and gentle pressure applied.

Injections in tortoises are generally given into the quadriceps. The area should always be cleaned with an antiseptic, prior to injection.

American box tortoises are somewhat harder to examine. If they clamp up, the finger and thumb can be pressed against the posterior carapace and plastron. This allows the anterior plastron to open slightly from the carapace. The finger and thumb are carefully inserted into this gap, and gentle pressure applied to open the hinge (Fig. 13.4). The front leg is then brought forward, in a similar way to that described for the tortoise. Thus the front legs are used for keeping the hinge open. Once the front legs are out, the head of the American box tortoise usually comes out by itself (Fig. 13.5).

In dealing with the red eared terrapins (*Chrysemys* spp) it is possible to examine the head by working the fingers under the underlying plastron towards the head, putting them behind the front feet and then pushing the front feet towards the head, so that the head is not allowed to turn round and bite. Examination of the head end of *Chelydra* spp and *Trionyx* spp without chemical restraint is not recommended.

Special Considerations

Dealing with young Chelonia, even *Chelydra* spp is easier than dealing with the adults, as they seldom inflict a painful bite. Care must be taken with all young chelonians, not to squeeze too hard

FIG. 13.3 Preparing a tortoise for examination of the head. (1) A tortoise clamped up with its front legs in front of its head. (2) Removing a leg by applying gentle pressure under the toes and pushing the foot towards the mid-line. (3) Pulling the leg out to place it against the side of the carapace, thus allowing the head to be seen. (4) Head on view showing the legs held against the shell and the head exposed. (5) Gently grasping the head on either side of the tympanic membrane and extracting the head from within the shell.

on the plastron or carapace, as this may damage the underlying developing bones. Similarly, more care must be taken when trying to extract the head or limbs from within the shell.

In general, sick tortoises and terrapins are easier to handle. They tend not to resist examination of the head, or withdrawal of a leg from the shell.

Transportation

Tortoises can be carried in zip-up bags, with newspaper at the bottom or in a cardboard box. If they are in a cardboard box, it should be large enough to let them move around, but not too large or they are likely to slide from side to side during transportation.

When dealing with the aquatic species, it is advisable that they are kept moist. It is often difficult to carry them with adequate amounts of water. If a terrapin or turtle is in a tank, or a bucket, with a large amount of water, it is more likely to be damaged than if it is in a box with a soaked towel underneath and on top of it.

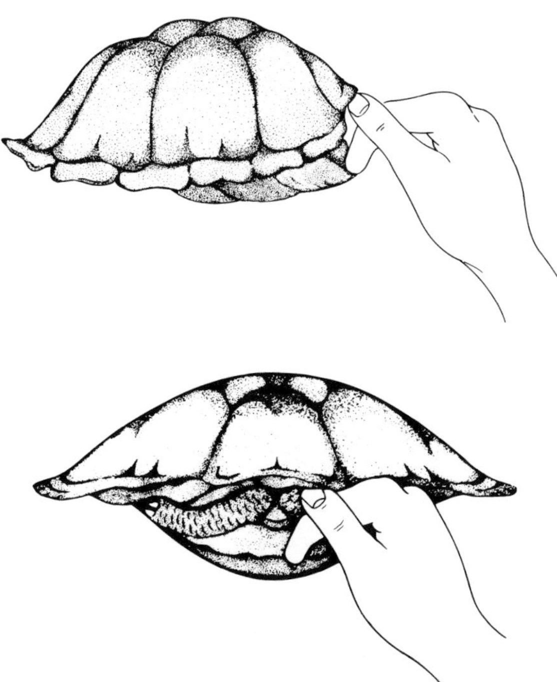

FIG. 13.4 Cross-finger action of opening the 'box' of an American box tortoise.

Snakes

Snakes are members of the Order Squamata (Sub-order Serpentes), together with lizards. It is widely held that it is easy to distinguish snakes from lizards by the fact that snakes do not have limbs, whereas lizards do. This is not entirely true, as there are also legless lizards (slow worms). There are also members of the snake family, particularly the boids (e.g. boas and pythons) which have rudimentary hind limbs, seen as vestigials or spurs on either side of the cloaca.

The correct way to distinguish between a snake and a lizard is by the following:

1. Snakes do no have ear drums, whereas lizards do.
2. Lizards have fully functional eyelids, whereas in snakes the upper and lower eyelid are fused together to form a transparent membrane, known as the spectacle.

Snakes consist of a large number of species, some of which are covered in the United Kingdom by the Dangerous Wild Animals Act 1967. There are 2400 species of snakes, and they are divided into 11 families. The most common ones that will be encountered are the Boidae (pythons, boas), and the Colubridae (rat snakes, garter snakes, corn snakes). The Elapidae and Viperidae are poisonous groups of snakes and are thus included under the Dangerous Wild Animals Act 1967 in the United Kingdom.

Approach and Handling

Snakes' mouths are formidable weapons. Their multiple sharp teeth can give a nasty bite. Snakes have backward-pointing teeth so that, once bitten, if you try and withdraw your hand quickly from the mouth, you will just make the wound worse by tearing the skin.

All snakes must be approached with caution. They are fast movers and can constrict as well as bite, therefore larger snakes, e.g. Indian pythons (*Python molurus*), reticulated pythons (*Python reticulatus*) and boa constrictors (*Boa constrictor* subspp.), should only be handled with someone else present to assist if necessary. A large python constricting round an arm can prevent the use of that hand.

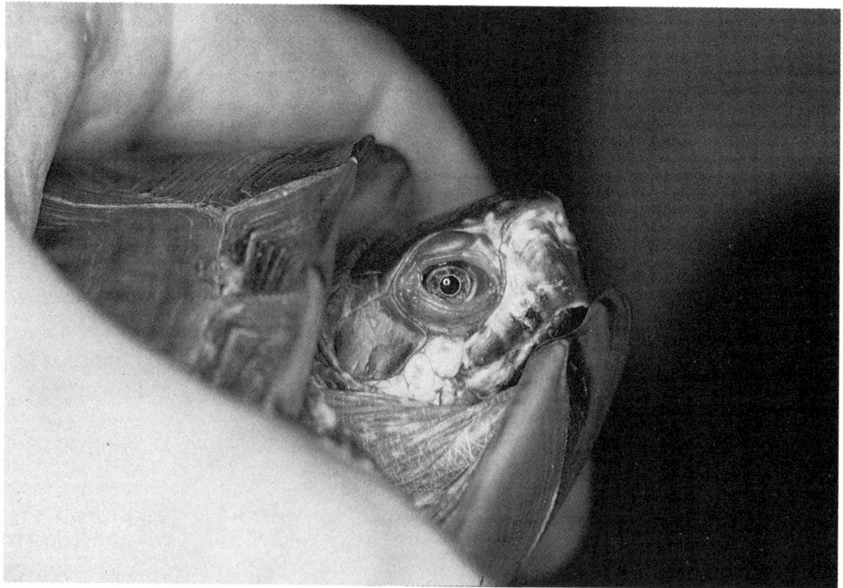

FIG. 13.5 An American box tortoise held with its front legs against the sides of the carapace so that the head appears.

When dealing with a snake in a snake bag, first open the bag and look in to see where the snake is. If you can see the head, it is advisable to reach for it from the outside of the bag, so that the head is grasped using the bag as a protective device between the handler and the snake's head (Fig. 13.6). Once the head is held, put the other hand into the bag and grasp the neck just behind the head. The snake may then be taken out from the bag with one hand holding the head and the other

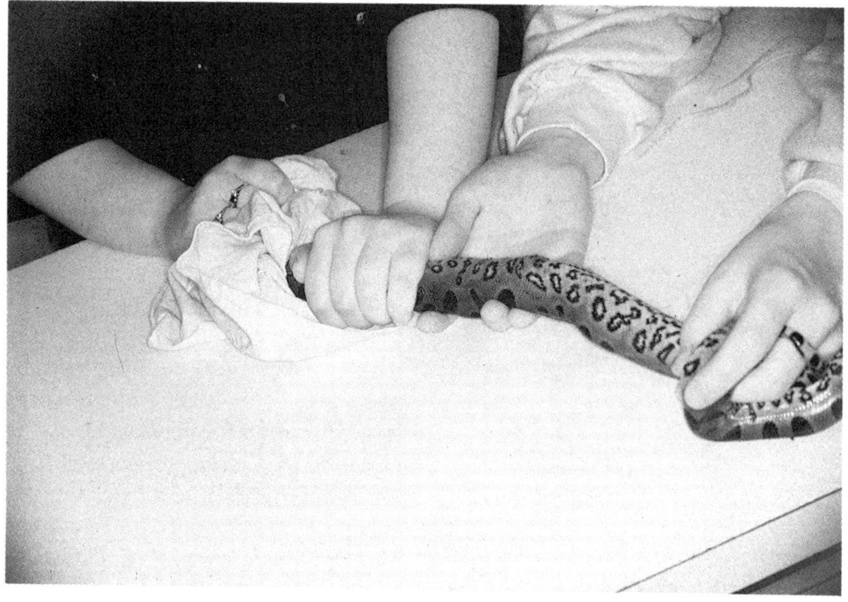

FIG. 13.6 An anaconda grasped through a bag, as it is a 'bad tempered' snake.

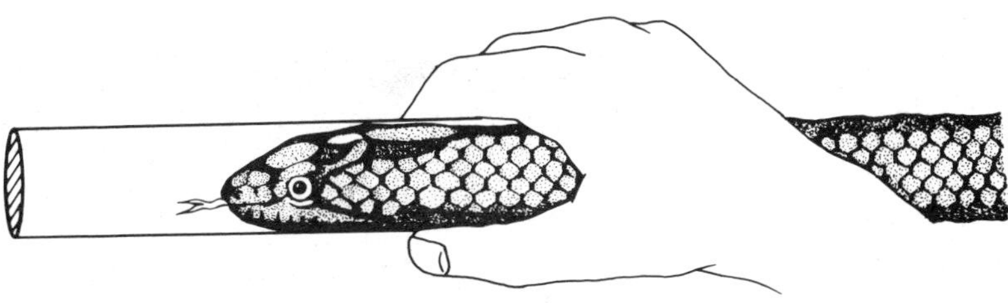

Fig. 13.7 A snake in a sealed tube gripped at the opening of the tube so that the head end is inside.

hand holding the middle of the body. As far as possible, try not to grip a snake tightly, but hold it gently and firmly. If it is known to be a friendly snake, then the head may not need to be restrained. It is possible to let the snake move freely from hand to hand. Excessive force when gripping a snake may result in death, not immediately, but five to ten days later, due to toxaemia from the autolysis of damaged muscles. Gentle movements towards snakes are recommended. A fast movement is more likely to result in the snake striking.

Dealing with Vicious or Poisonous Snakes

It is not advisable for an inexperienced person to deal with poisonous snakes. However, some vicious snakes, such as reticulated and Indian pythons, may well have to be handled. The general guide line on dealing with vicious snakes is similar to that for poisonous snakes.

Snakes always tend to go for a darkened hole, so one way of capturing them is to place a darkened tube, which has been blocked at one end, near the cage. As the snake goes into the tube it is possible to grasp the opening of the tube and the body of the snake. The tube has to be such that the snake cannot turn round in it (Fig. 13.7).

Another technique is to use the snake stick. This is a Y-shaped stick, with a rubber inner tube, stretched across the top of the Y. A vicious snake may be pinned by its head and neck to the ground. The hand is placed around the base of the neck and the head, with a finger on top of the head (Fig. 13.8), before it is picked up. The difficulty is not necessarily picking up a poisonous snake, but safely putting it down afterwards. The way to put it down is to press the finger so that the head is pressed against the tank, and then push down as

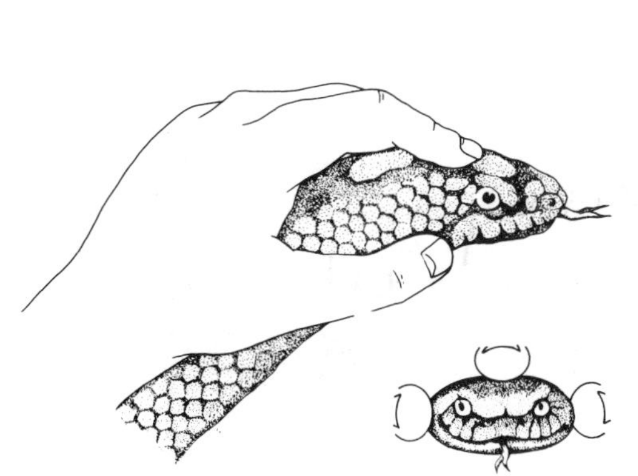

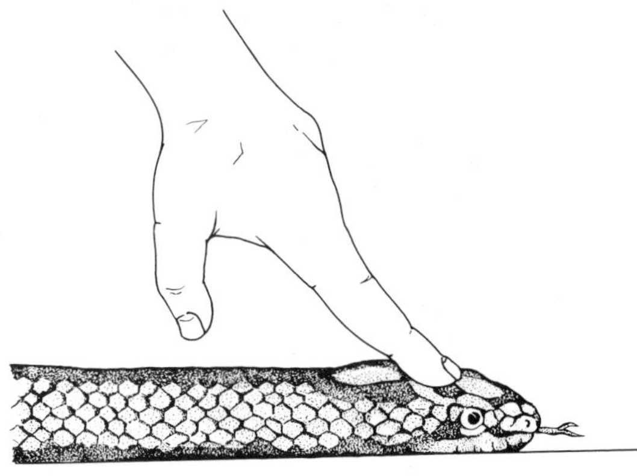

Fig. 13.8 *(left)* Holding a poisonous snake, with the thumb and third finger at about the occiput level and the index finger on top of the head (*see inset for head-on view*). *(right)* Releasing a dangerous snake, by pushing against the ground with the index finger and then quickly withdrawing the hand.

the hand is taken away very fast. Vicious or poisonous snakes may also be slowed down by cooling, prior to attempted handling. If detailed examination is required, it is advisable to use chemical restraint.

Special Considerations

When dealing with very young snakes, it is important that the person going to handle them is not a smoker, as a young snake's skin is very permeable to nicotine from the finger. This may poison the snake and result in death. Sick and dehydrated snakes must be handled with care. Just handling and stressing a dehydrated snake may well result in it going into shock and dying.

Transportation

It is advisable to transport a snake in a bag. This can either be a specially made bag, a canvas money bag, or a pillow case. For the very large species, a continental quilt cover can even be used. The end of the bag is folded over and tied or knotted. The snake in the bag should then be placed in a suitable container, to stop it from getting out or damaging itself during transportation.

Lizards

Lizards belong to the sub-order Sauria and have roughly 3750 species. They are divided into 16 families. The ones commonly encountered will be the Gekkonidae (geckos), the Iguanidae (iguanas), the Skinkidae (skinks), and the Lacertidae (green lizards).

All lizards must be handled with care. They have sharp teeth, which can give a severe bite, and some, especially Iguanidae, also have talons that can scratch an unwary handler.

Lizards must never be grabbed by the tail. They are able to demonstrate autotomy and shed an appendage, resulting in the lizard in one hand and the tail in the other which will not impress the owner. Although a tail that is lost will regrow, it will never grow back to its former beauty.

Approach and Capture

The majority of lizards are amenable towards being handled, especially if they are pets, in which case they are probably quite used to it. The exception is probably the Tokay gecko (*Gekko gekko*), which tends to be a very aggressive individual. Special care must be taken with this creature, as it also has stickers on the pads of its toes, which will allow it to walk up walls, glass and along the ceiling. If it gets out and escapes, it is therefore particularly hard to recapture.

The ideal way to handle the larger lizards is using two hands, one round the pectoral girdle, and the other around the pelvic girdle, clasping the limbs against the tail to hold it still and to stop it from struggling. For larger iguanas, it might be advisable to wear gloves when doing this procedure, to avoid being scratched (Fig. 13.10).

Another little trick, especially when one is needing to keep a lizard still, is to place a thick blanket, or towel, over its head. Once it has its head covered, it will quite often remain motionless, so that its body may be examined. Some species of lizard, e.g. the water dragon (*Physignathus* spp), will remain immobilized when turned on their backs, with a finger placed over the third eyelid.

Transportation

Lizards can be transported in an enclosed container, similar to the plastic pet carriers that are used for cats, or in a large polystyrene container, which should be closed so that the lizard cannot easily get out.

Crocodilia

All crocodilians are covered by the Dangerous Wild Animals Act 1976 in the United Kingdom. Care must be taken when handling these species, especially the larger ones, as they can be very dangerous. Apart from the fact that they have very sharp teeth which, in an adult, are capable of removing a limb, they also can use their claws as talons, and scratch. In addition, they can use their tail as a whip and unbalance and knock a person over. Do not be deceived, some crocodilians can move very fast when they want to!

Handling

Smaller crocodilians can be picked up by placing one hand very quickly behind the head, to grip this

FIG. 13.9 Restraining a small iguana with a hand around the pectoral girdle and limbs.

and the pectoral limbs. The other hand holds the pelvic limbs against the tail similar to that already described for the lizard. However, it is advisable that suitable gauntlets are used. For extra safety, it is possible to bandage the jaws together. In the larger species a towel may be used.

Larger species require a lot more experience in handling. The ideal method is to fall on top of

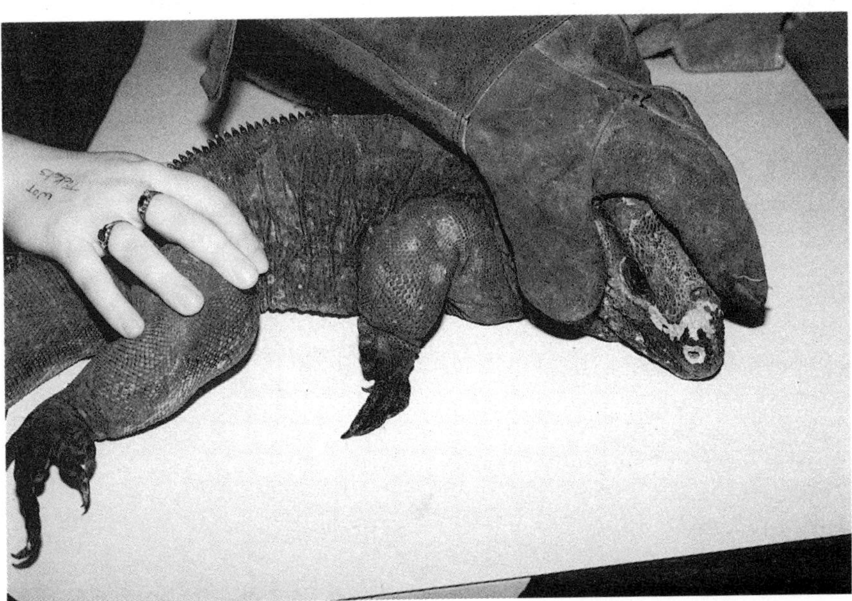

FIG. 13.10 Handling a rhinoceros iguana with a glove.

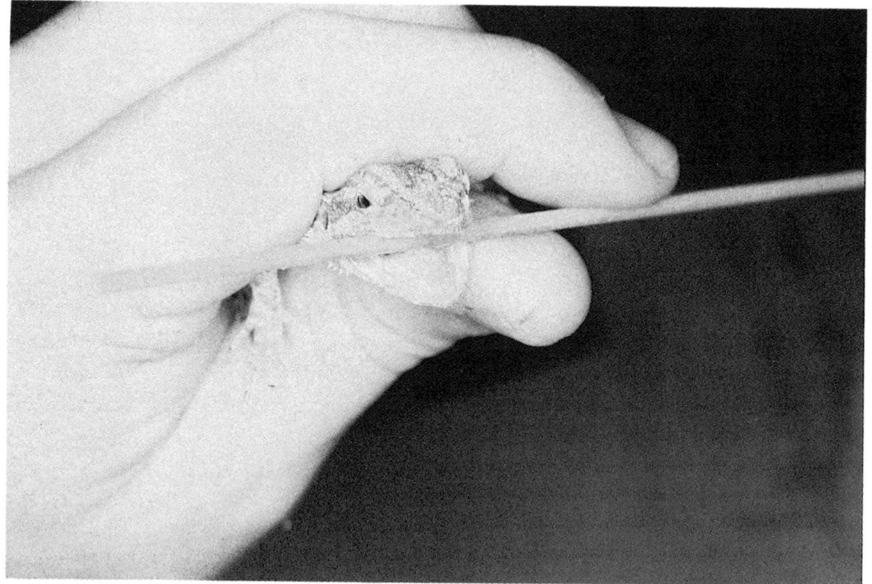

FIG. 13.11 A small wooden spatula may be used to open a lizard's mouth for examination.

them, putting the hand around the neck whilst gripping with the back legs around the tail. However, with larger species, handling should only be attempted by an experienced handler, or with chemical restraint. Younger crocodiles, although looking harmless, must still be treated with respect as they may inflict a painful bite.

Transportation

Crocodilians should be placed in a box or wooden crate of a suitable size so that they are reasonably confined and not knocked about during transportation, or able to move excessively. They need not be in water, but it is important they are either sprayed regularly, or kept in a moist environment, using, for example, a soaked towel.

References

Frye, F. L. (1981) *Biochemical and Surgical Aspects of Captive Reptile Husbandry*. Krieger, Melbourne.

Jacobsen, E. R. and Kollias, G. V. (1988) *Exotic Animals*. Churchill Livingstone, London.

Cooper, J. E. and Jackson, O. F. (1981) *Diseases of Reptilia* Volume 1 and 2. Academic Press.

Fowler, M. E. (1986) *Zoo and Wild Animal Medicine* 2nd Edition. W. B. Saunders, Philadelphia,

Frye, F. L. (1979) Reptile Medicine and Husbandry in the Veterinary Clinics of North America, *Symposium on Non-Domestic Pet Medicine*. Editor Bover W. J. W. B. Saunders, Philadelphia.

Jackson, O. F. (1976) In *The Care and Treatment of Children's Exotic Pets*. Editor Cowie A. F. BSAVA Publications, Cheltenham, pp 19–37.

Lawton, M. P. C. (1989) *Handling and Examination of Reptiles*. Video Produced by Unit For Continuing Veterinary Education, Royal Veterinary College, University of London.

Obst, F. J., Richter, K., and Jacob, U. (1988) *The Completely Illustrated Atlas of Reptiles and Amphibians for the Terrarium*. T. F. H. Publications Inc. Neptune City, New Jersey.

Further Reading

Cooper, J. E. and Beynon, P. M. (eds) (1991) *Manual of Exotic Pets*, 2nd edition. BSAVA Publications, Cheltenham.

14

Farmed Fish

LYDIA BROWN

Introduction

Farmed fish live in aquatic environments which vary widely from coldwater species (e.g. rainbow trout) living in fresh water, to marine species which prefer warmer temperatures (e.g. eels). Some fish need different conditions of water quality throughout their life in order to mature (e.g. Atlantic salmon). All species of farmed fish have a common physiological make-up in that they are poikilothermic and their basal metabolic rate varies with ambient water temperature. This means that at high water temperatures the metabolic rate of fish is elevated, whereas the converse is true at lower water temperatures. On farms, fish are reared either in tanks, raceways, net pens or ponds. They are often held under very intensive conditions (stocking densities may be as high as 36 kg/m^3). One of the most common tasks performed on fish farms is netting and grading animals, so that equivalent sizes of fish share the same tank. (Salmonids are carnivores and will eat smaller fish if frequent grading is not carried out.)

Another feature of farmed fish is that they react adversely to stress. Any minor disturbance may constitute a stressor for fish, for example crowding, elevated water temperatures, reduction in dissolved oxygen or even overfeeding. Plasma cortisol concentrations in salmonids remain elevated for a long time after a stressor has been introduced to a system. Thus, the constant need for handling farmed fish throughout the production cycle, combined with their extreme physiological response to stressors' means that the handling and restraint of farmed fish must be undertaken with care.

Methods of Approach and Capture which Allow General Control to be Established for Most Procedures

Methods of approach vary with the facility in which fish are kept and so each system of holding farmed fish will be dealt with separately. However, some basic procedures apply for all these facilities. Movements around fish-holding areas should be made smoothly and quietly with the minimum of fuss. All preparations for netting the fish should be made in advance. Water temperatures need to be at a comfortable level for the fish and there should be a high concentration of dissolved oxygen in the water before any handling procedures are embarked upon. Fish should be starved for twelve to twenty-four hours before handling and sound (vibrational) stimuli should be kept to a minimum. If light can be controlled (e.g. by operating in a room where the light has been dimmed) then so much the better.

The type of handnet used to capture the fish should receive special care. It should be made of knotless nylon or rope to avoid trauma to the skin of the fish. The handnet should be shallow so that only a few fish are netted at one time. The net should have a wide mouth. Rectangular, as opposed to circular-mouthed nets are preferred (Fig. 14.1). The net must be placed in the water gently so as not to stress the fish at the moment of capture.

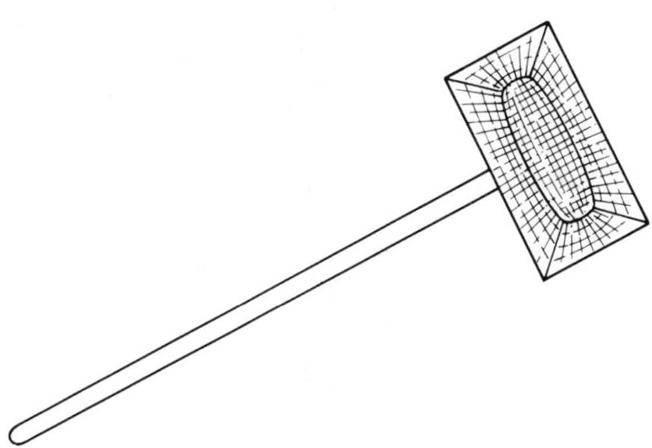

FIG. 14.1 An ideal handnet to catch fish.

Tanks and Raceways

The level of the water in the tank or raceway should be lowered before attempting to catch fish. It is generally less stressful to crowd fish in a smaller volume of water before placing a hand net in the water to catch them. Water levels should be reduced gradually. Any overhanging automatic feeders should be removed to allow easy access to the fish.

Earth Ponds

In some cases it is possible to lower the water level, but even if this is not possible, the approach to capturing fish is the same. Fish may be seined from one end of the pond to another (Fig. 14.2). The seine net will be weighted at the bottom end and two people generally walk, either in the pond or on the bank, pulling the net along-side them across the shortest side of the pond. The fish should not be crowded too quickly. If they panic, the fish can jump out of the water and over the net, or else swim underneath the leadline at the bottom of the net. The floats at the top of the seine net should never disappear below the surface of the water. If they do, the fish will escape the net by swimming over it. This procedure can be used as a grading mechanism whereby the mesh of the net is chosen to let smaller fish escape through the mesh, leaving larger fish to be harvested or moved to another growing pond. In very large extensive lakes, seine nets may be used by anchoring one end on the bank and using a boat to pull the net around the lake or pond in a wide arc.

FIG. 14.2 Schematic diagram to show how to seine a Danish earth pond.

Sea Cages/Net Pens

Fish held in the sea are generally held in floating cages or net pens. When attempting to capture fish, the sides of the cage are raised up bringing them towards the surface. The net can then be manipulated so that all the fish are in one corner. The principles for handnetting in earthponds and net pens are the same as for raceways.

Common Manipulations Used in Handling the Species

All of the manipulations described below will cause some stress to fish. Chemical methods of tranquillization and anaesthesia are available and should be used for fish when long term manipulations are required. Fish can be handled in a variety of ways and the manipulations are based on the following:

Using Equipment

Handnets are generally used to manipulate fish from one tank to another. They should not be used to hold fish for long periods. This is especially important in air, since the reflex movement the fish makes when it tries to escape will cause severe trauma to the skin resulting in epidermal sloughing and loss of scales. In the long term, this may lead to ulceration and bacterial infections. Also fish can be held out of water in a damp cloth (e.g. Hessian bag).

Fish have no eyelids and are therefore extremely photophobic. Consequently, if the damp cloth is relatively heavy and covers the whole body, especially the eyes, the fish will be more quiescent than if left uncovered. Gentle manual pressure on the head and tail will restrain fish adequately for many minor procedures.

Manual Restraint

Fish can be held (using wet hands or gloves) for short periods by grasping the caudal peduncle (the 'wrist' of the tail) with one hand and cupping the head of the fish with the other hand. Some salmonid farmers will walk several yards with fish simply by holding the caudal peduncle, although this is not advised.

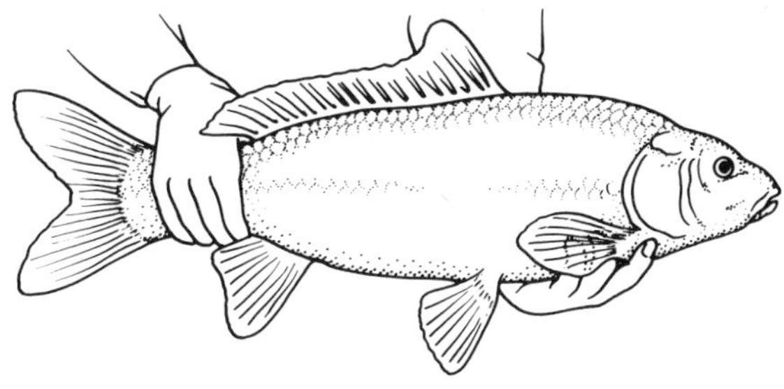

FIG. 14.3 Manual restraint of a fish.

Experienced cyprinid farmers can hold broodstock and walk several hundred metres with them by placing the left hand under the abdomen with fingers clasped around the pectoral fins. The right hand then holds the caudal peduncle (Fig. 14.3). The farmer can move with the fish as it twists and turns without letting go. However, this is an accomplished art and is generally only done satisfactorily by experienced fish keepers.

Lower Ambient Temperatures

In some Scandinavian countries fish farmers may place brood fish in the snow, covering them up and leaving them for several minutes before they perform any procedure (such as stripping the fish). The lower ambient temperature has the effect of reducing the basal metabolic rate of the fish thus causing the fish to react more sluggishly to external stimuli.

Special Considerations which Relate to Aged, Very Young, Sick, or Unhandled Individuals

All farmed fish should be considered worthy of special consideration in that they never completely adapt to the stress of handling, crowding and grading on a farm. However, some species of fish show particular susceptibilities to these procedures at specific stages of their life.

When Atlantic salmon smoltify and are taken from freshwater to seawater, they are especially vulnerable to stress, and must be handled very cautiously at this delicate stage of their life cycle.

Fish should be starved prior to being transported, and water temperatures in the transport chamber should be lowered. The fish will become inappetent for about a week after being moved so, although food should be offered, it should be done so at reduced rates (e.g. 0.25 per cent body weight per day instead of 3 per cent body weight per day, according to species and time of year).

Where possible, diseased fish should never be handled while they are sick. All attempts should be made to keep environmental conditions constant. Fish should not be crowded and water flows should be increased, thus helping to reduce the stocking density.

Handling as it Relates to Transport

Fish are transported by road (in trucks), by air (using helicopters), or by sea (using deep-well boats and barges). Farmed fish must always be starved for twelve to twenty-four hours before transport. If they regurgitate during transport, the excess nitrogenous compounds in the water, coupled with the stress of transport, may kill all the fish. They should be hand-netted in small numbers into the transporter as described earlier. The tank must have constant aeration supplied to it throughout the journey and have a means of measuring dissolved oxygen and temperature at any time.

The water in the transport facility may be cooled by placing chipped ice on the surface (in order to reduce the basal metabolic rate of the fish). Once at their destination, fish ought to be handled as little as possible. It is preferable to let the fish drain with the transport tank water into the new facility through a dump drain on the tank, rather than net them out of the tank into the new facility.

Transport of farmed fish must only occur when environmental conditions (e.g. water temperature, dissolved oxygen and total ammonia nitrogen concentrations) are favourable to the fish. If these basic principles are not adhered to, large mortalities may occur. Attempts to transport salmon smolts in the sea at the height of summer from cage to cage can result in 75–80 per cent mortalities.

Further Reading

Huet, M. (1979) *Textbook of Fish Culture*, Fishing News Books Ltd, Surrey, England.

Roberts, R. J. (1989) *Fish Pathology*, Baillière Tindall, London.

Sedgewick, S. D. (1973) *Trout Farming Handbook*, Seeley, Service and Co., London.

15

Ornamental Fish

DAVID M. FORD

Introduction

The first rule of handling live ornamental fish is — don't! Fish are cold-blooded animals and the human hand will 'burn' off the mucus layer leaving the fish with no protection from the fungus *Saprolegnia* which is always present in mature waters. The author has seen fungus growing on previously handled large fish in the exact shape of the fingers of the handler. With a surgical glove, a large fish can be handled underwater. A small fish can be handled through a fine mesh net. Anaesthetized fish should be laid on a towel soaked in the fish's water.

Examination of a sick fish can take place in four locations: the home aquarium, a pond, in a jar brought to the veterinary premises, or in an inspection tank. The latter is preferred. To have an inspection tank ready to hold any specimen brought by a client is good practice, it makes the diagnosis easier and shows professional expertise.

The Inspection Tank

There are three groups of aquarium fish: coldwater (e.g. small goldfish), tropical (e.g. community fish such as Angelfish and Neon Tetras) and marine (e.g. coralfish). Each type requires different water conditions. Aquarium goldfish will however, accept tropical water, and so the inspection tanks can be of just two types — freshwater and saltwater.

The most widely used tank is the all-glass, silicone-sealed type measuring 60 cm × 30 cm × 30 cm (24 in × 12 in × 12 in) and this is the easiest and cheapest model to obtain (Fig. 15.1). Some types have plastic surrounds which act as a buffer at the base. If this is absent, use a polystyrene buffer (ceiling tiles are ideal) between the bench and the base of the tank.

A cover is essential as fish placed in strange waters attempt to jump out. It can be a sheet of glass (3 mm, ground edges for safety) or plastic (acrylic is best, polythene will bow). An angle-poise lamp is probably the easiest source of lighting and illumination can be changed from overhead to side as required.

No gravel or ornaments are needed or indeed, are desirable. However, if a fish is to be left in the tank for any length of time, a weighted plastic plant will help the fish orient itself and so reduce stress. A sheet of glass or rigid plastic within the tank can be used to waft the fish forward to a viewing area. This is far better than the technique usually recommended which consists of trapping a fish for inspection between a fish net and the front glass. Nets totally stress fish, to the point of death by heart attack in some delicate species.

The freshwater aquarium should be heated to 24°C using a heater-thermostat. The cheapest and most reliable type of heater-thermostat is the combined unit held in a glass tube. If preferred, separate thermostatic controls are available, the lastest models using solid-state controls. An aquarium digital, stick-on thermometer, though inexpensive, is sufficiently accurate. The saltwater aquarium should be at about 26°C, which reflects the temperature of the Coral Reef. Many marinists keep coral fish at temperatures up to 30°C which enhances both the colour and the activity of the fishes. However, higher temperatures lower the oxygen content and good aeration is thus essential. A specimen brought to the veterinarian with thermal packaging will have cooled just a few degrees. The polybag should be floated to equalize the temperature or, if a bucket is used, partial water changes between the tank and container over 5 to 10 minutes will equalize the temperature gradually. Sudden changes in temperature stress the fish.

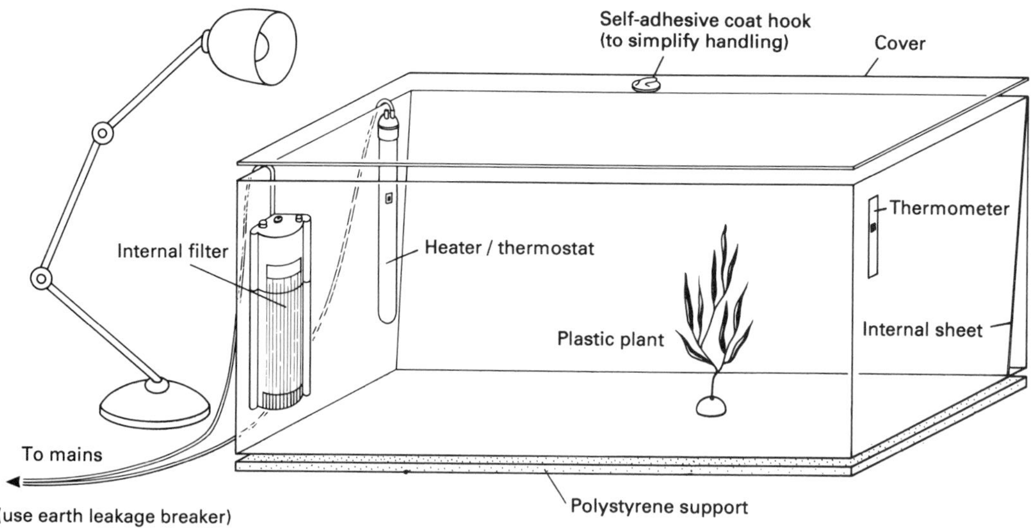

FIG. 15.1 The inspection tank: an aquarium with minimum life-supporting accessories.

The Fresh Water Tank

The inspection tank cannot be just filled with tapwater. The differences between the mature transit water from the aquarium and the 'raw' fresh water will result in chemical shock which may perhaps kill a sick fish. The water in the inspection tank must be aged and mature, i.e. contain nitrifying bacteria (*Nitrobacter* and *Nitrosomonas* spp). To achieve this, a filtration system must be included, which causes the oxygenated water to flow over a filter medium on which *Nitrobacter* convert ammonia and nitrite, excreted by the fish, into nitrate.

The simplest filter system for the 60 cm inspection tank is an internal power filter. This, like the many other commercial models available in the hobby trade, uses a foam filter. Water is drawn through the foam via a mains electricity motor sealed in a thermal resin for safety. The drive is coupled to a plastic impeller by a magnetic coupling. An effective improvement for the inspection tank with its low stocking density, is to replace the foam with ceramic pieces. This improves the biological filtration activity, which is more important than the additional mechanical filtration needed in the busy aquarium.

Since, for much of the time, there will be no fish in the inspection tank, nitrifying bacteria will need nutrients. These can be supplied by crumbling flaked fish food into the water daily; six days out of seven is sufficient if more convenient. If the flaked fish food is used, just three flakes, crushed and wetted out, will suffice. Monitor the maturing process by measuring the nitrite content (aquarium kits based on colorimetry are cheap and accurate) which should pass through a crisis value over a few days. When the nitrite value, (which peaks at about 10 mg/l NO^{2-}) has fallen back to zero, the water is mature and the filter operating biologically. Adding fish for inspection will not add chemical shock to their transit stress. The water will also be clear and clean for viewing the fish.

The remaining apparatus, so beloved by aquarists, such as airpumps, undergravel filters and special lighting, are not needed. The internal powerfilter will oxygenate the water sufficiently via an outflow near the surface.

The Salt Water Tank

It depends on the location of the veterinarian's practice as to whether a saltwater fish aquarium is needed. The most popular pet (in the world) is the common goldfish and this will probably be the most frequent species brought to the hospital.

Next is the larger tropical fish, such as the Oscar or Piranha, which has become the family pet. Marine fishkeeping has remained a low percentage of the hobby because of its complexity and high cost. However, that same high cost of individual fish makes it more likely that an owner will consult a veterinarian.

The same set-up as used for the freshwater inspection tank can be installed for this occasional customer. If the practice is in an area with several marinists and the traders who supply them, then a regular supply of sick fish may need to be catered for and a larger filter unit will be needed to maintain the zero nitrite reading essential for marine fish. An external powerfilter, such as the Atlantis X500, with a (unfiltered) turnover of about 500 litres/hour can be used (anything larger will give too much turbulence in the aquarium). The filter medium can be ceramic or carbon pieces. If carbon is used, trap the pieces in a bag (a nylon stocking is ideal) as the pieces float and can foul the impeller. The usual polyester fibre barrier recommended for the home aquarium filter is not really needed.

The water needs to be made saline to a specific gravity of 1.020 at 26°C. This is a low value to maximize the oxygen content. The initial pH will be 8.5 when proprietary seasalts are used. To monitor conditions, a plastic hydrometer is available to measure specific gravity and a pH paper is sufficiently accurate for the alkalinity reading. The biological activity of converting nitrite to nitrate releases hydrogen ions so that the pH tends to fall with age. Any value below pH 8 will stress coral fish, so water should be buffered with sodium carbonate to at least pH 8.2. Partial seawater changes will also restore the alkalinity.

The same feeding regime such as that used in the fresh water tank, for example, tropical flake foods, can be applied. There are bacterial cultures and nutrients available to speed-up the maturing process. It is important to maintain the marine tank at zero nitrite as the hypotonic saltwater fish are very susceptible to poisoning.

Maintenance

Once established, a filtered inspection tank should only need a strip-down cleaning annually. Beware of excess light however, because this will result in problems with algae. Green water is not harmful to fish, indeed it is beneficial, but it interferes with observation. A polyfoil barrier on the sides, or as a cover, will stop photosynthesis. Once the nitrite value is zero, cut back on feeding if it increases again. The best maintenance is by way of partial water changes. For a 25 per cent water change, water straight from the tap can be used. For 50 per cent or more, a dechlorinator such as sodium thiosulphate solution, should be used prior to addition to the aquarium. Proprietary dechlorinators are available. Saltwater will need adjusting to the correct specific gravity but partial changes should only be infrequent.

The filter must be maintained biologically active. To do this, strip it down for rinsing in just tepid tapwater. Do not sterilize or replace the filter medium and its valuable colony of nitrifying bacteria. Cross infection/infestation must be within the veterinarian's judgement. Most fish parasites are obligatory and if a few days pass without reuse of the inspection tank, any parasites such as *Ichthyophonus* or *Saprolegnia* in the freshwater and *Oodinium* or *Cryptocaryon* in the saltwater will have perished.

Diagnosis of contagious diseases such as tuberculosis (a zoonosis) may require a clean-down. Sterilization of nets and other aquarium equipment with proprietary disinfectants is best effected by iodo-compounds. However, iodophors (and especially chloro-compounds) are very poisonous to fish, so rinsing must be thorough. Other safe disinfectants include potassium permanganate, hydrogen peroxide, methylene blue and malachite green. Isopropyl alcohol sprayed twice on the emptied tank's inner surface will sterilize the corners effectively. Where possible, the filter should be cleaned without sterilization, but if it is necessary, the 'nitrite crisis' ageing process must be repeated.

Examination

A delivered fish will almost certainly have faecal matter in the transit bag or jar. This can be filtered off for microscopic examination after the fish itself is allowed to swim from the container into the holding tank. If the temperature difference is gross (i.e. detectable by hand) floating the container in the tank will allow equalization. However, a

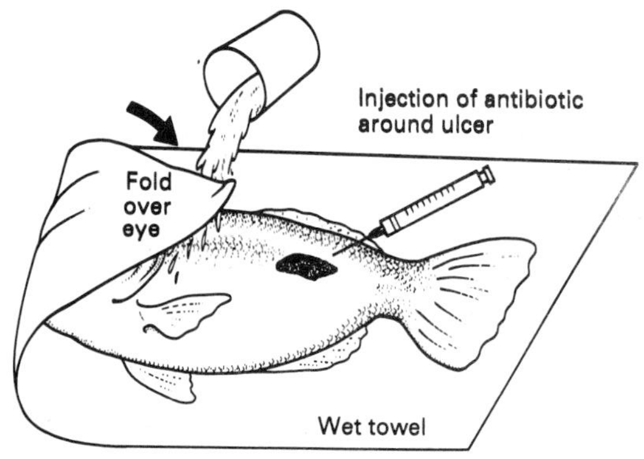

FIG. 15.2 An anaesthetized fish should be kept wet, especially the gills, during treatment.

chilled fish recovers quickly in a warm aquarium, it is rapid falls in temperature which are stressful.

Two hand lenses are needed to view the fish. A simple reading type of ×3 magnification is effective at about 8 cm to view the swimming fish, brought forward by the internal panel. A ×10 or more hand lens can be used at 2 cm on anaesthetized fish lifted from the aquarium.

The gloved hand can be placed in the aquarium to tilt the fish to check 'horizontal eye movement syndrome'. Large, non-aggressive fish can also be cupped for a mucus scraping to check for ectoparasites. Small fish, large active fish and certain poisonous fish (such as the Lionfish, *Pterois volitans*) should be anaesthetized. Anaesthetics should be given in a separate container, so the fish can be returned to the inspection tank for recovery.

There are currently eight notifiable diseases in fishes but the only one that may be found in ornamental fish is Spring Viraemia of Carp (SVC) which affects pondfish such as Koi. SVC is caused by the virus *Rhabdovirus carpio* and there is no effective treatment, so stock needs to be destroyed. MAFF publishes a description of the disease with addresses for notification in the United Kingdom.

The Home Aquarium

Visiting the home aquarium has the advantage of inspecting husbandry methods. Most ornamental fish diseases are a consequence of bad practice, e.g. from overcrowding, lack of filtration, unsuitable species, inappropriate temperature or water chemistry. Few fish are totally free of parasites and it is the excess mucus on the body or in the gills, caused by poor husbandry, that triggers outbreaks of excessive parasitic infestation. All aquarium waters carry a high bacterial load and a fish needs to be without stress to cope with this situation. Treating one affected fish in the hospital may not be the solution to the problem.

A visiting veterinarian should have a portable inspection tank (a 30 cm × 20 cm × 20 cm plastic tank, for example) and a siphon tube to fill it with the aquarium water. This can also be used for anaesthesia. A test kit for freshwater nitrite levels is essential. The marinist will have his own kit for testing the seawater and be familiar with the findings, but it is worth checking that such kits are reading accurately.

A home visit can also include showing the aquarist how to administer antibiotics, if prescribed. It also allows the veterinarian to assess biological load, and perhaps even weigh the fish. It is important to remember that mammalian doses of antibiotics are fatal to fish. It is always wise to refer to the literature and not to rely on memory for dose levels.

The Pondfish

The stately Japanese Carp called 'Koi' (the Japanese word for 'love') are the popular fish for garden ponds. Some owners have elaborate and very expensive filter systems to give clear water in their ponds for viewing the fish. Koi of pedigree quality can change hands for very large sums which makes veterinary attention essential. Owners will not usually risk catching and transporting such valuable fish to the hospital. In these cases the veterinarian should include a white baby bath, a large polythene bag, and a pond net in his equipment. The net is not used to capture the fish but to coax the sick Koi (or a large goldfish) into the bath. Great care is needed to ensure that everything and everybody does not also land up in the pond. If the fish needs to be lifted out for further treatment, it is best scooped into a large polythene bag. This can also be used for anaesthetizing the fish before lifting it out on to towelling soaked in the pond water. A section of the towel is draped over the fish's eye and pond water is occasionally trickled

Fig. 15.3 Pondfish are scooped into a plastic bag prior to anaesthetising.

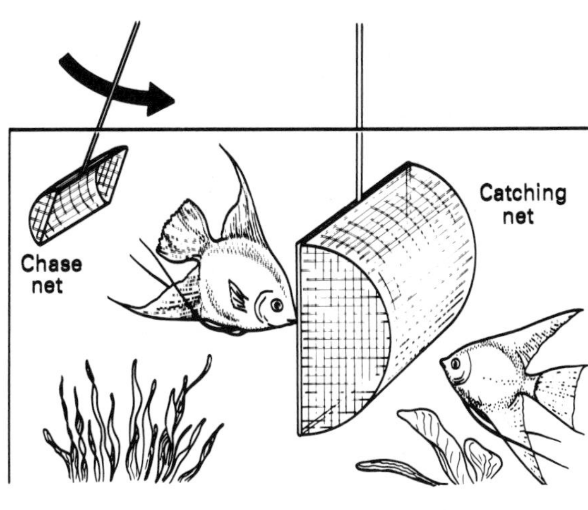

Fig. 15.4 Using two nets to catch fish quickly.

over the gills (Fig. 15.2). Wounds can then be cleaned and injections made, before returning the fish for recovery in the floating bath.

If bathing the fish in a chemical treatment is necessary, the same baby bath can be pressed into service. Dosing ponds should be avoided in all but extreme cases. The water volume is too great for cost-effective treatment and subsequent dilution is difficult. The filtration system of the Koi pond may be damaged by chemotherapy with disastrous consequences for a shoal of prize Koi.

Capture and Transporting of Fish

If a net must be used to capture a fish, always use two in a pincer action so that the fish takes flight from one net and will dive into the other (Fig. 15.4). Chasing a fish from one end of the tank to the other with one net gives great stress, even death. A green net is preferred, the open white net must look like the mouth of a shark to the fish, triggering primitive fear and flight reactions. Also, choose a fine mesh net as it is less likely to become entangled in the fins, barbels or scales. For the rougher skin of marine fish, it is best to line the net with (punctured) polythene. Large or dangerous fish should not be netted but enticed into a bowl or plastic bag, or anaesthetized *in situ* (Fig. 15.5).

A plastic bag, one third filled with water from the aquarium or pond, is used to move the fish (Fig. 15.6). The top two thirds can be air, caught and trapped by sealing with an elastic band, but for journeys of several hours it is better to use oxygen. Small fish can be bagged together but beware of panic swimming into the bag corners, where the gills can be closed by the plastic sheet, suffocating the fish. Round-bottomed bags are best, but square bags can be used by tying-off the corners with small elastic bands. A plastic bucket may seem the obvious choice for transporting fish, but the slopping of the water (especially during car journeys) dashes the fish against the sides causing bruising.

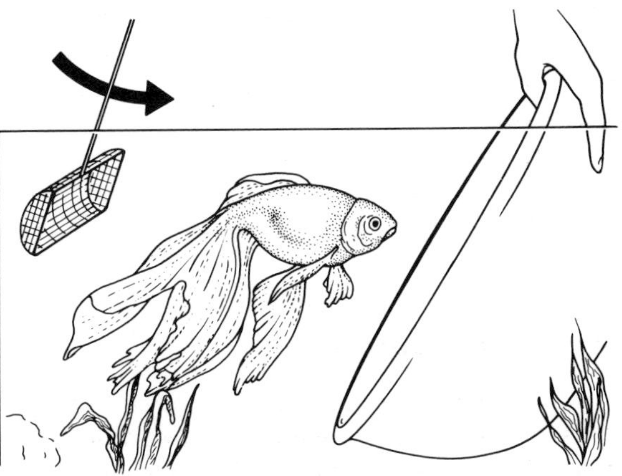

Fig. 15.5 Larger fish are scooped up into a bowl.

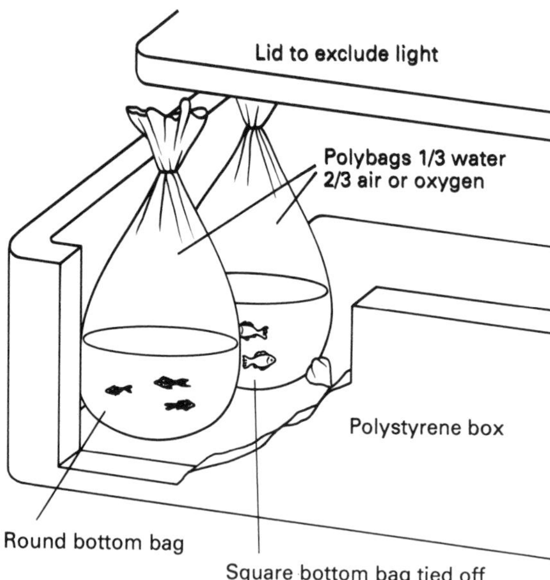

FIG. 15.6 Transporting fish; tropical and marine fish must be kept warm and given sufficient oxygen.

The temperature needs to be maintained for tropical and marine species. A polystyrene box (as used for importing fish, or even a picnic box) is ideal, but a carrier bag will suffice if lined with paper or woollens. Light must be excluded to reduce the fish's metabolic rate during the journey. Coldwater fish should be wrapped to exclude light.

Reference

MAFF Publication UL122 BL5866 (1989) from Lion House, Willowburn Estate, Alnwick, Northumberland, NE66 2PF.

Further Reading

Amlacher, E. (1970) *Textbook of Fish Diseases*, TFH Publication No. PS-667, New Jersey.

Andrews, C., Exell, A. and Carrington, N. (1988) *The Interpet Manual of Fish Health.* Salamander Books Ltd, London.

Carrington, N. (1985) *A Fishkeeper's Guide to Maintaining a Healthy Aquarium*, Salamander Books Ltd, London.

Cooper, J. E., Hutchinson, M. F., Jackson, O. F. and Maurice, R. J. (1985) *Manual of Exotic Pets*, BSAVA, Cheltenham.

van Duijn, C. (1973) *Diseases of Fishes*, Iliffe Books, London.

Elkan, E. and Reichenbach-Kline, H.-H. (1974) *Colour Atlas of the Disease of Fishes, Amphibians and Reptiles* TFH Publication No. H-948, New Jersey.

Ford, D. M. (1987) *Successful Fishkeeping* and *Fishkeeping Made Easy*. From Aquarian Advisory Service, P.O. Box 67, Elland, W. Yorks. HX5 0SJ.

Hoffman, G. L. and Meyer, F. P. (1974) *Parasites of Freshwater Fishes*, TFH Publication No. PS-208, New Jersey.

Reichenbach-Kline, H.-H. (1973) *Fish Pathology*. TFH Publication No. PS-204, New Jersey.

Reichenback-Kline, H.-H. (1977) *All about Marine Aquarium Fish Diseases*. TFH Publication No. PS-747. New Jersey.

Untergasser, D. (1989) *Handbook of Fishes*, TFH Publication No. TS-123, New Jersey.

16

Small Mammals

PAUL FLECKNELL

Introduction

Handling of small mammals presents a number of difficulties. These can easily be overcome providing that time is spent learning the correct approach to these animals. Although the risk of serious physical injury to the handler is relatively low in comparison with the risks associated with handling large animals such as horses, small mammals can still inflict painful bites. The majority of rodents and rabbits maintained either as pets or in research institutes are relatively docile, and few problems should be encountered provided that appropriate handling techniques are employed. A different approach is required when handling wild mammals, and this is discussed in more detail at the end of the chapter.

Initial Capture and Restraint

Small Rodents

The initial capture of domesticated small rodents is made considerably easier if they are confined to a small transport box or cage. Attempts at capture when the animal is surrounded by exercise wheels, food hoppers, water bowls and similar items often result in unnecessary stress to the animal. Practising veterinary surgeons should instruct clients to bring small rodents to the practice in a shoe box or similar container. It is important to remember, however, that small rodents may gnaw through a cardboard container in a relatively short time, so the box should not be left unattended for more than a few minutes. Rodents housed in research institutes will usually be maintained in cages which allow good access to the occupant, in contrast to the narrow trap door arrangements commonly seen on pet rodent cages. Before opening the cage or transport container it is advisable to close any doors or windows in case the animal should escape.

Small rodents are generally easy to restrain, but their small size makes them especially vulnerable to physical injury, not least by the handler inadvertently dropping them when bitten. Although generally non-aggressive when accustomed to handling, they have extremely sharp incisor teeth which can inflict a painful bite. If you are bitten by a small rodent and the animal retains its grip, do not attempt to pull it free or shake it off. This can result in a much more substantial wound being produced by the backward pointing incisor teeth, and if the animal is shaken free it may be seriously injured. The animal should be placed in its cage and released. It will almost invariably cease biting and relax its grip.

To establish initial restraint, a firm but gentle approach is required, as with any other animal species. The following methods are recommended when dealing with pet or laboratory animals.

Gerbils

Gerbils vary considerably in their response to handling, and may be extremely active and difficult to catch. Although they possess a long tail, which might seem to represent a convenient means of restraint, this should not be grasped because the skin is readily shed if subjected to more than minimal traction. If the animal is accustomed to being handled, it can be cupped in the palm of one hand. More secure restraint is achieved by first immobilizing the animal by cupping in both hands, or by placing one hand over the animal. The fingers can then be positioned so as to surround the animal's body, with its head protruding between thumb and forefinger (Fig. 16.1). Alternatively, the skin overlying the back and neck can be gripped as described for the mouse (see below).

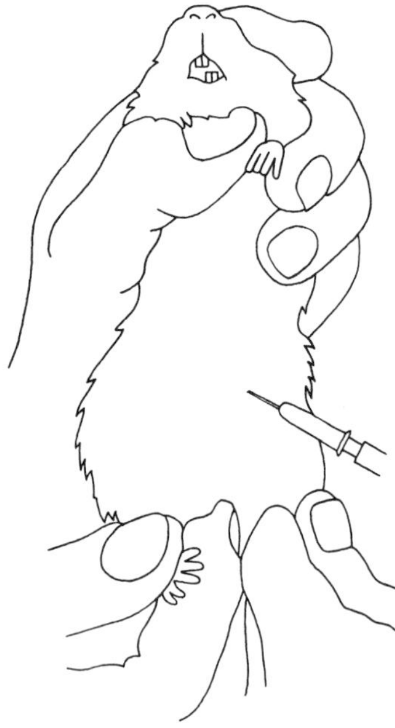

FIG. 16.1 Restraint of the gerbil to enable intraperitoneal injection. The injection should be carried out by an assistant.

Guinea Pigs

Guinea pigs are docile animals but they are easily alarmed by inexpert attempts at restraint. When frightened, the animals will move rapidly around their cage and will then be very difficult to catch. To avoid this problem, approach the animal rapidly and smoothly, and grasp it around the shoulders with one hand (Fig. 16.2). It can then be lifted clear of its cage. It is essential that only minimal pressure is placed on the thorax, and that restraint is applied primarily with the thumb, first and second fingers around the shoulders. If the guinea pig weighs more than 200–300 g, it is best to provide additional support to the hindquarters by cupping these with the other hand (Fig. 16.3). This is essential when handling larger animals (>1000 g) or pregnant females. The thumb should be positioned under the mandible, or the forelegs can be held so that they cross beneath the chin, so that the animal cannot lower its head to bite if some minor manipulation is to be undertaken (see later).

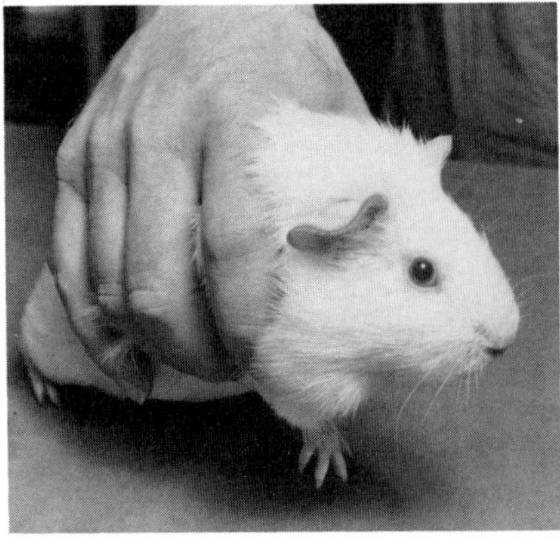

FIG. 16.2 Initial restraint of the guinea pig is achieved by grasping it around the shoulders.

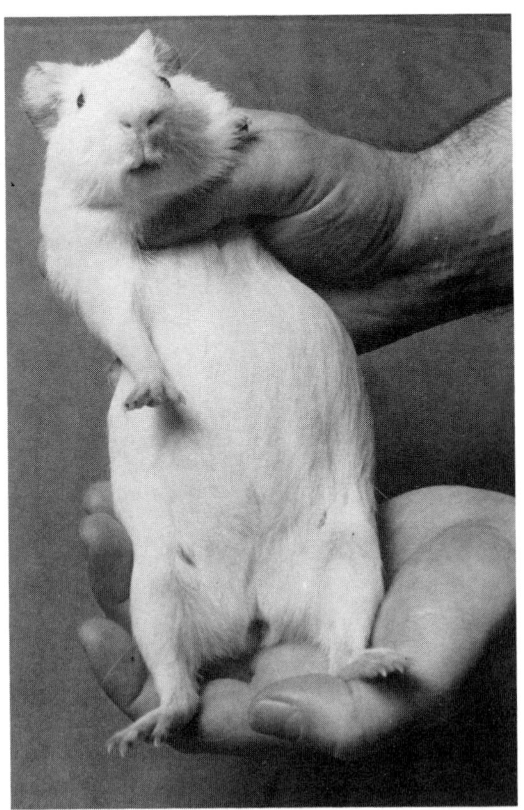

FIG. 16.3 Restraint of the guinea pig for examination; the hindquarters should be supported if the animal weighs more than 200–300 g.

Hamsters

Hamsters can be restrained relatively easily by cupping in both hands, followed by grasping by the scruff. If the animal appears aggressive, it should

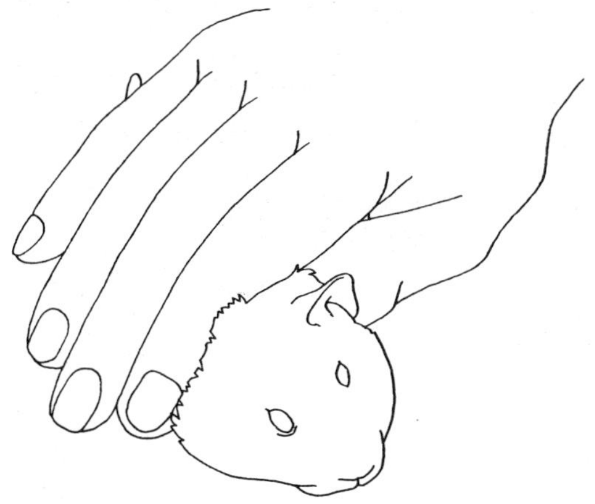

Fig. 16.4 Initial restraint of the hamster, by covering the animal beneath the palm of one hand.

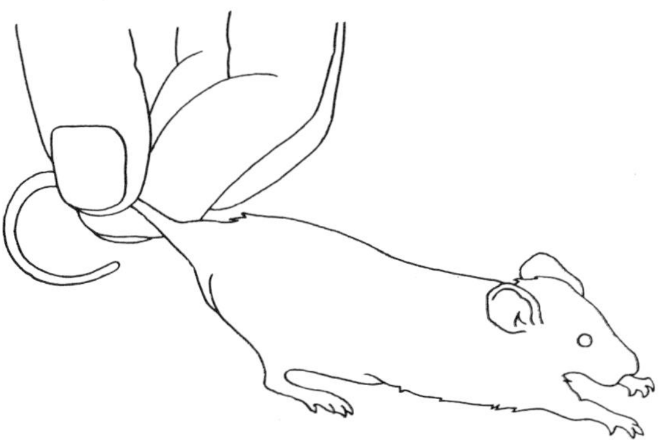

Fig. 16.6 Restraint of the mouse. The tail should be grasped by its base, the animal lifted clear of its cage and placed on a rough surface.

first be immobilized by covering its body with one hand, leaving the head exposed (Fig. 16.4). The skin overlying the neck and thorax can then be drawn up between the handler's fingers and thumb to provide a firm grip of the very extensive scruff (Fig. 16.5). The animal can be then be lifted clear of its cage and examined without risk of being bitten. At no stage of the examination procedure should the animal be allowed to run unrestrained on the examination table. Most hamsters appear unable to perceive the drop from the table edge to the floor and may fall and injure themselves.

Fig. 16.5 After immobilizing the animal as shown in Fig. 16.4, the scruff can be grasped between the handler's fingers and thumb.

Mice

Many mice will attempt to bite when restrained and a careful approach is necessary. The animal should be grasped by the tail, preferably the proximal third, and lifted clear of its cage. It should then be placed on a rough surface such as a piece of towelling, or allowed to rest on the sleeve of the handler's laboratory coat. If gentle traction is maintained on the tail, the animal will grip the towelling and attempt to pull away (Fig. 16.6). An initial examination of the animal can be carried out with the mouse in this position. If additional restraint is required, then the scruff can be grasped between the thumb and forefinger whilst maintaining a grip on the tail (Fig. 16.7). The mouse can now be lifted clear of the towel and the grip on the tail transferred to between the third and fourth fingers (Fig. 16.8).

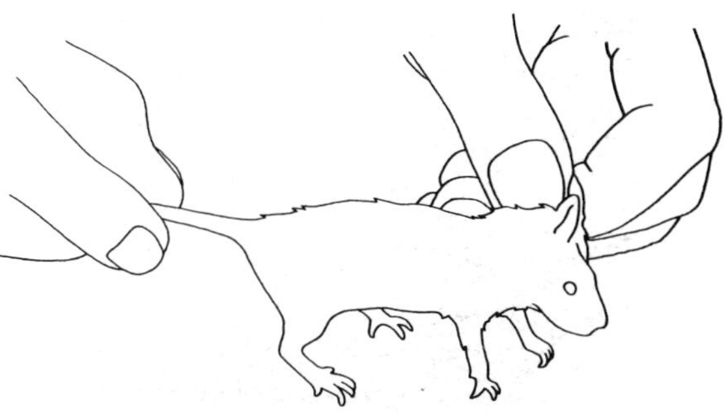

Fig. 16.7 After restraint as shown in Fig. 16.6, the animal's scruff can be grasped between the handler's thumb and forefinger.

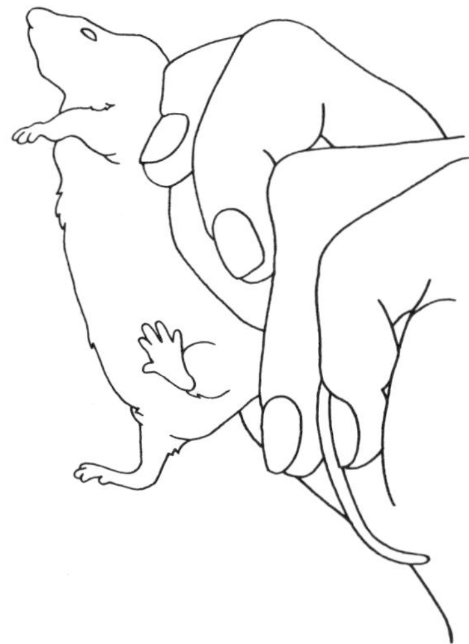

Fig. 16.8 Following restraint by the scruff and the tail as shown in Fig. 16.7, the mouse can be lifted and the grip on the tail transferred to between the handler's third and fourth fingers.

Fig. 16.9 Rats should be lifted from their cage by grasping around the shoulders.

Rats

Rats are generally docile rodents, provided that they are approached and handled in a gentle manner. The rat should be picked up by grasping it around the shoulders (Fig. 16.9) and be lifted clear of its cage. The handler's thumb should be positioned under the mandible to prevent the animal from biting (Fig. 16.10). It is important that the grip around the rat's shoulders is sufficient to prevent its escape, but that only minimal pressure is applied to the thorax. Inexperienced handlers frequently tighten their grip if the rat struggles and this results in a severe impairment of chest movement. The rat will become severely distressed and struggle even more violently, become cyanosed, and will almost invariably bite when released. If the rat struggles when first restrained, it can usually be calmed by applying only minimal restraint and allowing the animal to rest on the handler's sleeve. It can be allowed to move around, with intermittent restraint being applied around the shoulders or to the base of the tail when it becomes necessary to adjust the position of the rat or to prevent its escape. After a short time most rats will become more placid, and they may then be restrained more firmly for examination and to enable any of the procedures listed below to be carried out. If the rat appears aggressive, then it can be lifted by the base of the tail, transferred to a rough surface as described for the mouse and then grasped around the shoulders. It is rarely necessary to grasp a rat by the scruff, and this method of

Fig. 16.10 After lifting clear of the cage, as shown in Fig. 16.9, the handler's thumb should be positioned under the mandible to prevent the animal from biting.

restraint is often resented by the animal unless it is accustomed to the procedure.

Rabbits

Rabbits are especially susceptible to the effects of stress and are easily injured by careless handling. They are usually non-aggressive animals, but when frightened they may make violent attempts to escape. Occasionally an aggressive individual that attacks the handler and attempts to bite or scratch will be encountered. Such animals are generally easily controlled by applying the restraint techniques described below rapidly and firmly. Rabbits have powerful hind legs equipped with sharp claws which can produce painful scratches.

To restrain a rabbit, grasp it firmly by the skin overlying the neck. Under no circumstances should the animal be restrained by grasping its ears. Some handlers include the ears in their grasp when holding the scruff. This is acceptable provided that no tension is placed on the ears, but an owner or untrained person observing the procedure can easily misinterpret this means of restraint and assume that the rabbit is being lifted by its ears. For this reason it is recommended that the ears should not be grasped when restraining rabbits.

Following initial restraint by grasping the scruff, the handler's other hand should be placed around the hindquarters (Fig. 16.11). If the animal struggles, the handler's forearm can be positioned along the line of the back to prevent the animal moving. The rabbit can then be lifted clear of its box (Fig. 16.12) and positioned with its head tucked under the handler's arm and with its hindquarters and back supported (Fig. 16.13). These

FIG. 16.12 After restraining as shown in Fig. 16.11, the rabbit can be lifted clear of its cage.

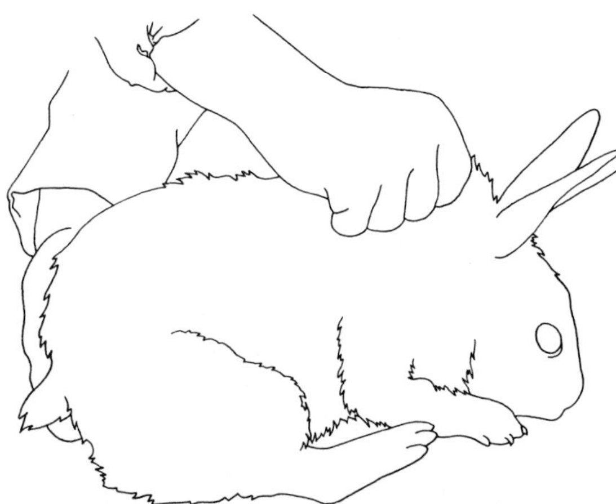

FIG. 16.11 Rabbits should be restrained by grasping the scruff, with the hindquarters restrained and supported.

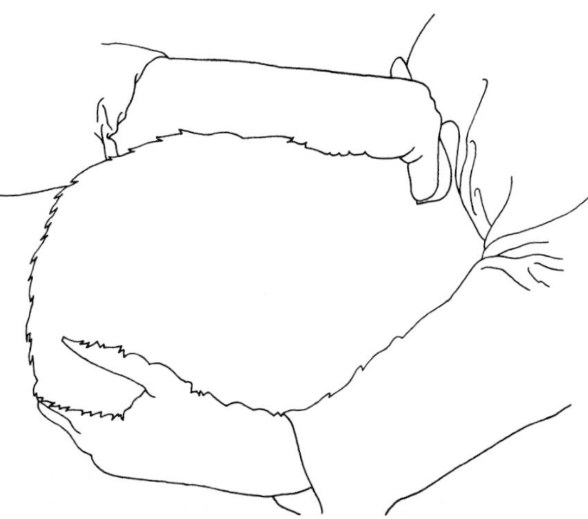

FIG. 16.13 Following removal from its cage, as shown in Fig. 16.12, the rabbit should be positioned with its head tucked under the handler's arm and with the back and hindquarters supported by the handler's forearms.

precautions are necessary to minimize the risk of the animal kicking out when it is lifted. If this occurs, the rabbit is likely to fracture or dislocate its lumbar spine, and may also injure the handler. It is also important to support the animal's back as securely as possible when placing it on an examination table. A common error made by inexperienced handlers is to release the animal before it is fully in contact with the ground. This can cause the animal to kick out and struggle, and once again can result in the animal injuring itself. When returning a rabbit to its cage or transport box, the handler should ensure that both his arms are withdrawn simultaneously, to avoid being scratched by the animal. When placing a rabbit in a cage, the animal should be positioned so that it faces one of the side walls. If the rabbit is released whilst facing the back of the cage and it kicks with its hind legs, then it could injure the handler, or send a shower of sawdust and faeces into his face.

Ferrets

Ferrets vary considerably in temperament, ranging from non-aggressive pet animals to more unpredictable animals which are maintained for rabbiting. Laboratory ferrets are normally reasonably easy to restrain, but occasionally a particularly aggressive animal may be encountered. The ferret should be grasped firmly around the neck (Fig. 16.14), and the hind limbs and pelvis gripped as the animal is lifted clear of its transport box or cage (Fig. 16.15). It is often helpful to distract the animal's attention before grasping it. Unlike other small mammals, the ferret has a powerful and well-muscled neck and this must be held firmly. The handler's thumb and forefinger should be positioned beneath the mandible so that the animal cannot lower its head to bite.

Fig. 16.14 Initial restraint of the ferret is achieved by grasping around the neck and shoulders.

Fig. 16.15 Following restraint as shown in Fig. 16.14, the animal can be lifted clear of its cage and the hindquarters restrained by holding it around the hind limbs and pelvis.

Common Manipulations

Gerbil

Following initial restraint as described above, the gerbil should be grasped around the shoulders and thorax, and additional restraint obtained by holding the hind limbs. An assistant should extend one of the animal's hind limbs, and an intraperitoneal injection can then be made into the posterior quadrant of the abdomen, with the needle being directed along the line of the hind limb (Fig. 16.1). Intramuscular injections should be made into the quadriceps muscle with the animal restrained in the same manner. Subcutaneous injections can be made into the flank. Alternatively, the method of restraint can be changed so that the scruff is grasped and the injection made into the anterior aspect of the skin overlying the neck. Intravenous injections can be made into the lateral tail veins but, because the skin of the tail is so fragile, some form of restraining device should be used, as illustrated for the mouse (Fig. 16.20).

It is often necessary to examine the teeth of small rodents and occasionally to clip their claws. The incisor teeth can be inspected with the gerbil restrained as for an intraperitoneal injection, but examination of the cheek teeth requires general anaesthesia. The same method of restraint will allow clipping of the claws, but this should be carried out by an assistant who can immobilize each limb in turn whilst clipping is carried out.

Hamster

Following initial restraint as described above, subcutaneous injection can be made into the extensive scruff, with the animal restrained as shown in Fig. 16.5. Once lifted from its cage, the hamster will usually open its mouth giving a clear view of the incisor teeth (Fig. 16.16). Inspection of the cheek teeth and cheek pouches is best carried out under general anaesthesia. With the animal restrained as shown in Fig. 16.16, an assistant can extend one hind limb and carry out an intraperitoneal injection into the posterior quadrant of the abdomen. The same method of restraint is suitable for carrying out intramuscular injection into the quadriceps. The hamster has no superficial veins which are suitable for intravenous injection in the conscious animal.

Guinea Pig

To enable an assistant to administer a subcutaneous injection into the scruff, the guinea pig should be restrained on a suitable surface with one hand positioned along either side of the animal. This method of restraint is also suitable for allowing a number of minor manipulations to be undertaken, for example examination of the incisor teeth, the ears, and the claws. Intravenous injection can be made into the ear veins, but the animal will usually shake its head during the procedure. Examination of the cheek teeth is best undertaken after the induction of general anaesthesia.

An alternative method of restraint for subcutaneous injection is to hold the animal as shown in Fig. 16.17. The injection can then be carried out by an assistant into the fold of skin in the inguinal region. Intraperitoneal injection is undertaken by the assistant with the animal restrained in a similar way. The assistant should grasp one hind limb and direct the needle along the line of the limb into the posterior quadrant of the abdomen (Fig. 16.17). Intramuscular injection can be made into the quadriceps muscle with the animal restrained as

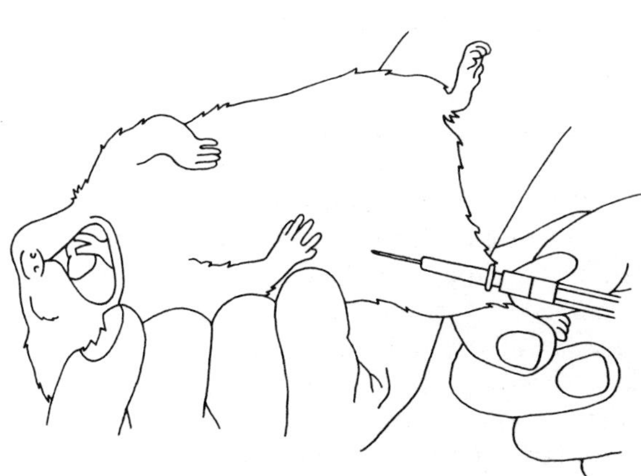

FIG. 16.16 Restraint of the hamster for intraperitoneal injection. An assistant should extend one hind leg and inject into the posterior quadrant of the abdomen.

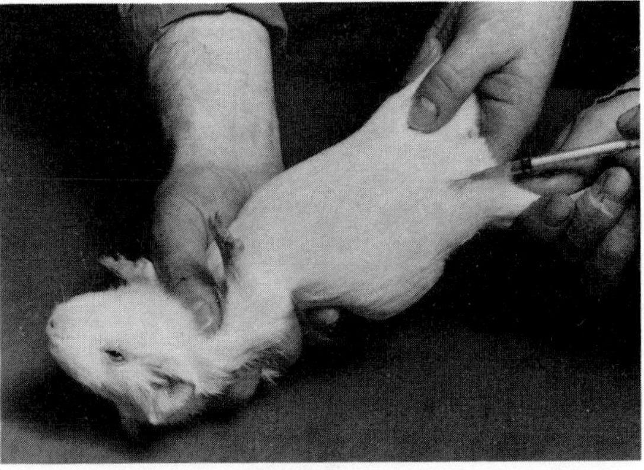

FIG. 16.17 Restraint of the guinea pig for intraperitoneal injection. The handler holds the animal around the shoulder and supports the pelvis and one hind leg. An assistant restrains the other hind leg and injects into the posterior quadrant of the abdomen.

FIG. 16.18 Intramuscular injection in the guinea pig is made into the quadriceps muscle. The assistant should immobilize one hind leg by grasping the quadriceps between finger and thumb.

shown in Fig. 16.18. An assistant should immobilize one hind leg by grasping the quadriceps between his forefinger and thumb, and the injection made into the middle of the muscle mass.

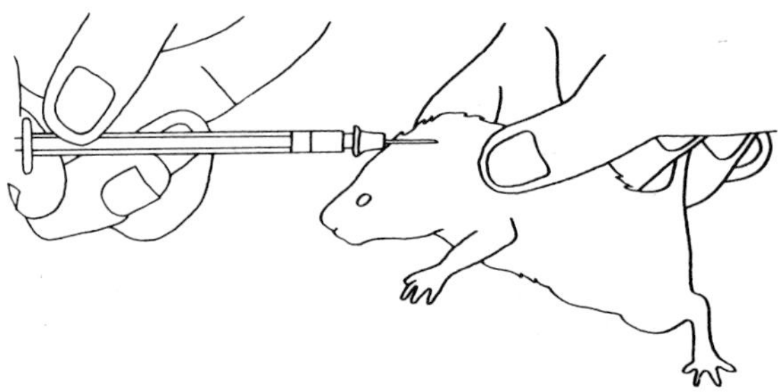

FIG. 16.19 Subcutaneous injection in the mouse. Care should be taken to direct the needle into the scruff and not into the handler's finger or thumb.

Mouse

The restraint technique shown in Fig. 16.8 is suitable for most examinations and manipulations. Intraperitoneal injection can be carried out as described for other species of rodent, although the small size of the mouse also enables single-handed injection into a posterior quadrant of the abdomen. Subcutaneous injection can be made into the scruff, with the mouse restrained as shown in Fig. 16.19, and intramuscular injection into the quadriceps. Both of these techniques require an assistant to administer the injection. Intravenous injection into the lateral tail vein is most easily undertaken with the mouse restrained in a purpose-made restraining device (Fig. 16.20).

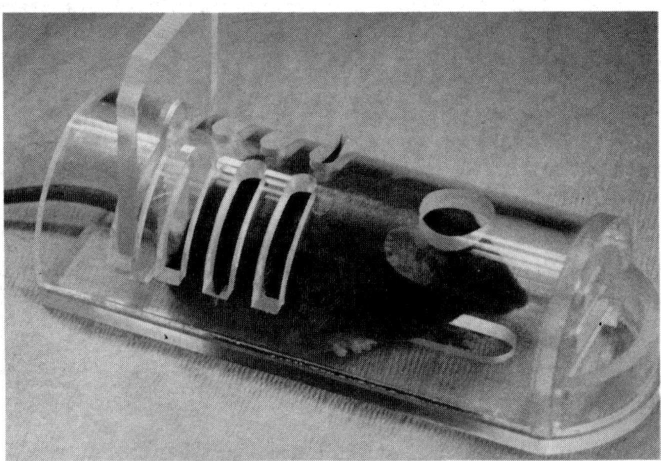

FIG. 16.20 A mouse restrained in one of the many different types of restraining devices that are available from commercial suppliers.

Rat

The method of restraint illustrated in Fig. 16.10 is suitable for most manipulations such as inspection of the incisor teeth and clipping of the claws. An assistant can administer an intraperitoneal injection by extending one hind limb and injecting along the line of the thigh into the posterior quadrant of the abdomen (Fig. 16.21). Intramuscular injection is made into the quadriceps following restraint of one hind limb by an assistant. Subcutaneous injections can be made into the flank. Alternatively the rat can be restrained by grasping around the shoulder and injection made by an assistant into the scruff.

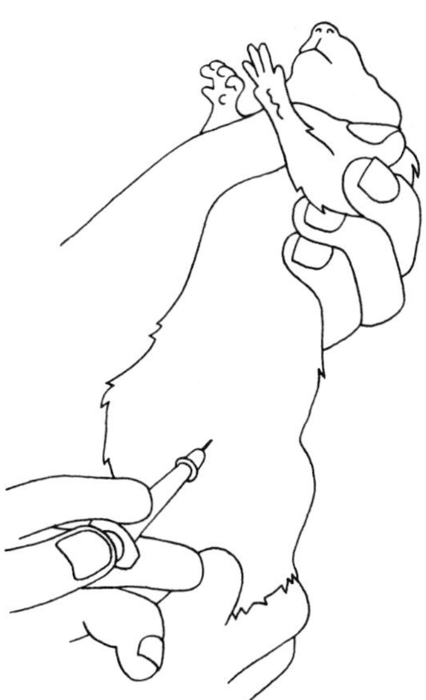

FIG. 16.21 Intraperitoneal injection in the rat. The handler restrains the rat around the shoulders and holds one hind limb. An assistant extends the other hind limb and injects into the posterior quadrant of the abdomen.

Rabbit

Intraperitoneal injection is best carried out by an assistant with the rabbit restrained as shown in Fig. 16.11. This method of restraint is also suitable for subcutaneous injections into the scruff. Although this procedure and intramuscular injection into the quadriceps can be carried out by the handler, it is preferable that the handler provides additional restraint by placing a hand around the animal's hindquarters, and allows an assistant to carry out the injections. With the animal restrained in this way, it is also practicable for an assistant to clip the animal's claws, to examine the incisor teeth, and to use an otoscope to examine the ears. An assistant can also administer an intravenous injection into the marginal ear vein. If an assistant is unavailable, the animal can be restrained by wrapping it in a towel or laboratory coat, with the head protruding (Fig. 16.22). Provided that the animal is securely wrapped, it will

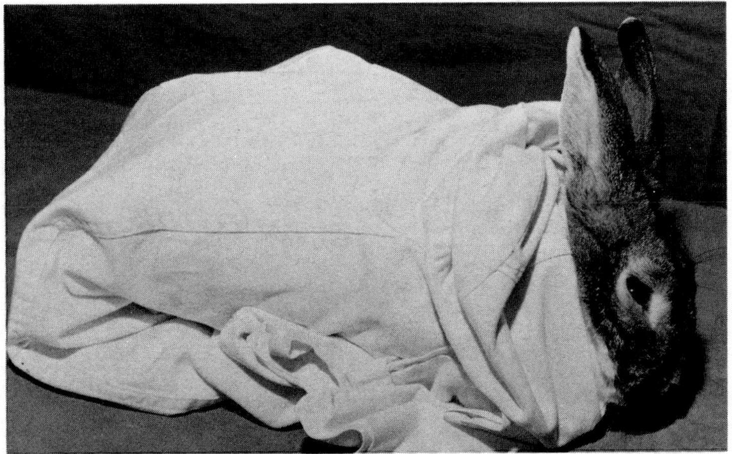

FIG. 16.22 Restraint of a rabbit using a laboratory coat. The rabbit is placed on the coat with its neck positioned over the collar. Both sides of the coat are wrapped securely across and around the rabbit. The bottom of the coat is then folded beneath the animal.

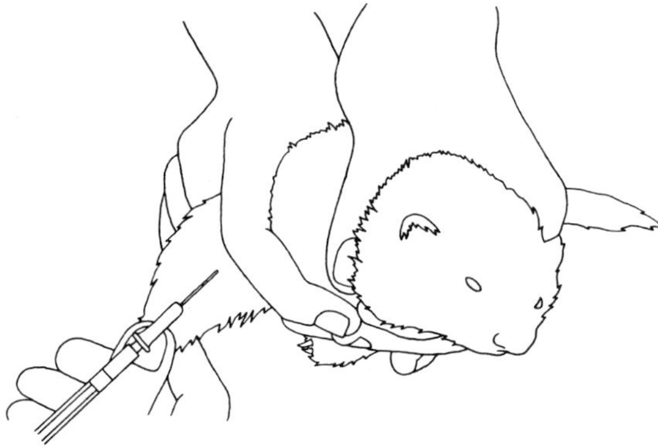

FIG. 16.23 Restraint of the ferret for intravenous injection. An assistant extends the forelimb and carries out the injection into the cephalic vein.

readily accept this method of restraint and will not struggle. Both of the operator's hands are then free to carry out an intravenous injection. Wrapping the animal in this way provides good restraint and supports the rabbit's back, and is generally preferable to using a rabbit restraining box. If one of these restraining devices is used, it must be of an appropriate size for the rabbit. If the device is fitted with a neck yoke, it is preferable to avoid using this attachment because of the risk of injury to the rabbit if it struggles whilst in the restraining device.

Ferret

Following restraint as described above, intraperitoneal injection can be made by an assistant into the posterior quadrant of the abdomen, as described for other species. Subcutaneous injection is made into the scruff, or into the loose skin of the flank, and intramuscular injections are made into the quadriceps. Intravenous injections can be made into the cephalic vein as shown in Fig. 16.23.

Special Considerations

Old Animals

Old animals may resent handling, especially if chronic painful lesions such as arthritis are present. Rats, guinea pigs and rabbits may become extremely obese when older, and the physical size of the animal may make restraint with one hand difficult. It may even be necessary to require an assistant to provide additional support when lifting the animal.

Older rodents, in particular rats, commonly have severe chronic respiratory disease, and the stress of handling and restraint can be hazardous to the animal. It is especially important that restraint is gentle and kept to the minimum necessary.

Young Animals

Handling of young rodents presents few problems associated with restraint, although juvenile rodents may be extremely active. Recently weaned mice, for example, may leap out of their cages if the lid is removed. Provided that this is anticipated they can be restrained rapidly, but if it is unexpected, the animal may escape and recapture can prove difficult.

Neonatal animals should be handled as little as possible, as mismothering following such disturbance is a frequent problem. It is advisable for the handler to wear surgical gloves, and to roll the neonates in their bedding after returning them to their cage to try to remove the scent of the handler. Neonatal rabbits are especially likely to be neglected following disturbance of the doe and handling of the young.

Wild Rodents

Wild rodents present some special problems with regard to handling. In contrast to pet or laboratory rodents, most wild rodents will try to bite the handler during restraint. For this reason, it is often advised that protective gauntlets are used. It is important to appreciate, however, that these animals will be highly stressed because of the fear associated with capture, and possibly because of physical injury caused during trapping. Rather than subject them to further physical manipulations, it may be preferable to avoid handling them. A suitable design of capture box and holding cage can enable transfer of the animals without the need for handling. If a detailed examination of the animal is necessary, then it may be preferable to anaesthetize it, for example by placing the capture box in an anaesthetic chamber and utilizing a volatile anaesthetic such as methoxyflurane.

Wild rodents may be infected with a number of

disease organisms which can affect man (zoonoses). It is therefore essential that special care is taken to avoid the risk of infection. Simple measures such as wearing gloves and adopting a high standard of personal hygiene may be sufficient to minimize the risk of contracting certain infections. In most instances, however, additional precautions will be needed. Caged rodents generate large quantities of dust and aerosols which may be inhaled by the handler. Minimizing this hazard may involve both the use of filtered face masks and negative pressure cabinets for holding the animals. It is also important to ensure that soiled bedding and cages are handled safely, and that the animals are housed securely.

References and Further Reading

Flecknell, P. A. (1987) *Laboratory Animal Anaesthesia*, Academic Press, London.

Harkness, J. E. and Wagner, J. E. (1989) *The Biology and Medicine of Rodents and Rabbits*. Lea and Febiger, Philadelphia.

Tuffery, A. A. (1987) *Laboratory Animals, An Introduction for New Experimenters*. Wiley, Chichester.

UFAW (1987) *The UFAW Handbook on the Care and Management of Laboratory Animals* (ed. T.B. Poole) 6th edn., Longman Group UK Ltd., Harlow.

17

Mink

PETER G. HAWKYARD

Introduction

The farming of mink began in America in the 1920s. In the last thirty years, however, production in Europe has increased and Denmark now leads the world, with an output of 12.5 million pelts per annum, approximately one-third of world production.

Mink are raised throughout the world under very similar conditions in family owned farms in which the mink are housed in open sided sheds. Down the middle of each shed is a broad walkway with a row of wire pens at each side, upon each of which is attached a nesting box. While there are some variations on this theme, they are not so great as to affect the principles of mink handling.

The handling of mink looks deceptively easy to the observer watching an expert, but it is not as simple as it looks. There is a knack to catching and handling a mink, which requires good reactions and manual dexterity. Starting with female mink, a sound technique can, however, be acquired by the novice in a relatively short time. Handling of fully adult male mink should not be attempted until reasonable proficiency with females has been attained, since males are twice the bodyweight (weighing on average 2.5 kg) and are much stronger than females of the same age.

Protective Gloves

Young mink (kitts) may be handled up to the age of six or seven weeks without protective gloves. Great care must, however, be taken to ensure that the mother mink has no access to the kitts. After this age, gloves are always used to handle mink. A great deal of thought has gone into the design of a mink handling glove. If the leather is soft, the mink can easily be handled because the leather is malleable, but the glove will be too easily penetrated by the mink's long, sharp canine teeth. If the leather is hard and/or thick, the handler will be well protected but will not have the necessary touch sensitivity needed for the handling of mink. The ideal glove is, therefore, a judicious compromise between one which provides good sensitivity but poor protection and one giving good protection but poor sensitivity.

Catching Penned Mink

The standard way to catch a mink which is in its pen and nestbox is to 'shoo' it from its nestbox into the pen. This can be done quite easily by either blowing on the mink and/or a quick shake of the nestbox, whereupon the mink will normally run into the pen. If a mink is particularly stubborn, the choices available to the handler are either to push the rear of the animal with an implement in the direction he wants the mink to go, or, if experienced, to open the nestbox and catch the mink within. (This latter option is not for the novice handler, as the movements of the mink are unpredictable and it may easily escape.) Once the mink is in the pen, the handler opens the pen door (which on most farms doubles up as the feed grid) and reaches towards the mink. If the left hand is used, the handler should reach down the left-hand side of the pen, whereas, if the right hand is used, then the approach should be down the right-hand side of the pen. If the pophole (box entrance) is always on the handler's right-hand side, it is worth mastering catching with the left hand, and, similarly, if the pophole is on the handler's left, then it is a smoother operation to catch with the right hand.

If the pophole is on the handler's right-hand side, and the mink to be caught is at the far end of the pen, the handler will put in his left gloved hand and reach towards the rear of the pen. The mink will invariably take the shortest direct route and run through the pophole. As the mink heads for the box, the handler grasps it quite firmly towards the rear of its body, allowing the mink to slip through until the tail is within the grasp. This can be gripped firmly by the gloved hand with little apparent distress to the mink. Gripping the tail, the handler pulls the mink from the nestbox, which by this time it will have entered, towards the rear of the pen, keeping the vertebral column straight. Once clear of the nestbox the mink is lifted out of the pen. The subsequent handling procedure depends on the purpose for which the mink has been removed from the pen, and also upon the character of the mink itself.

When handling pregnant females, extreme care must be taken, particularly when taking the initial hold of the mink. As the handler first grasps the body, then allows the mink to slip through his hand in order to grasp the tail, there must be no excessive pressure to the abdomen. The mink must be caught at the first or second attempt. Repeated attempts by an unskilful handler should be avoided at all costs to minimize damage to the mink.

FIG. 17.2 Handling an aggressive mink.

Carrying

The mink, which is now firmly held by its tail in one hand, will show a natural reaction to bend forwards and upwards and bite the glove. If the fore end of the mink is then placed on top of the nestbox, it will normally grip the surface with its forepaw and pull away from the handler. The other hand can then be placed under the mink's chest and the animal lifted clear (Fig. 17.1). Many mink may be happily carried about in such a manner without showing any signs of discomfort or stress. Some may, however, display a measure of aggression. These animals, when they are being held by the tail in the left hand and the forelimbs are on top of the nestbox, should be grasped by the right hand on the back of the neck and shoulders, then lifted clear with both hands (Fig. 17.2).

FIG. 17.1 Carrying a fairly placid mink.

Restraint

Vaccination

This method of restraint is used when presenting a mink for vaccination, holding the mink up well in front. With the mink firmly held by neck/shoulders and tail, the hand holding the tail should be repositioned to hold the rear feet against the gauntlet of the opposite glove. This protects the person using the syringe from being badly, and unnecessarily, scratched. Vaccinations are normally administered subcutaneously in the front leg.

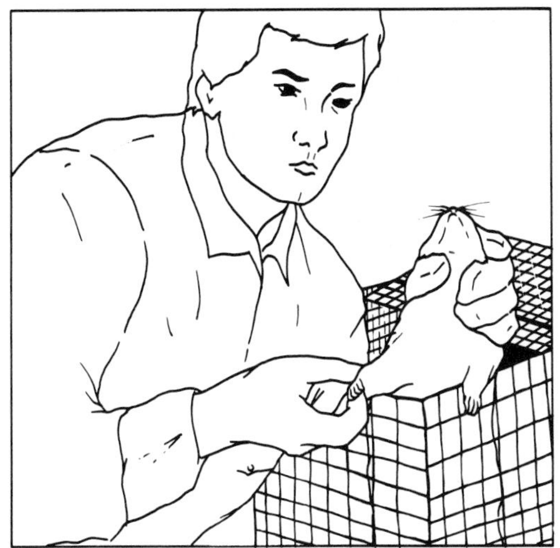

FIG. 17.3 Handling a mink in preparation for taking a blood sample.

Blood Sampling

Another exercise involving handling, which is regularly performed on a farm, is blood sampling. Sufficient blood to fill a three inch capillary tube is obtained by clipping the nail of a front paw just into the quick. This procedure involves one or two handlers and a sampler. Although there are several ways of restraining mink for this procedure, it is essential that the sampler has confidence in the handler and the method of restraint, since he or she is working barehanded. In the author's view, the safest way is to start with the mink held by the neck/shoulders in the left hand and the tail in the right, then to let go of the tail and, still maintaining a firm hold of the animal, put it back into the pen using the front right corner as a crush. The mink's head and neck are outside the pen, and the rest of its body is firmly constrained in the corner, evenly down the entire body, by the inside of the handler's left arm between the wrist and the elbow (Fig. 17.3). His right arm is free. Once the mink is securely held in this manner, the sampler can get hold of the front paw and do his job above the top line of the cage in reasonable safety. If the mink is particularly large and strong and the handler is unable to maintain a hold of the animal, it will more likely direct its efforts towards slipping back into the cage rather than escaping or biting.

Mother and Young

Most litters can be inspected and counted 24 hours after birth by running the hand, or dancing a piece of meat, on the pen top, thus tempting the mother out of the nest box. Once she has left the nestbox, a partition is put between pen and nestbox. This must be done quickly, as she will make strenuous attempts to return to her litter and it is absolutely vital to ensure that the partition is secure. Mink are excellent mothers, happily returning to their litters after handling. They will generally also accept kitts fostered from another litter. The latter practice is quite common amongst the mink farming industry in order to give kitts a better chance of survival. If they have been born in a litter of, say, eleven or more, the chances of their survival are greater if two or more are fostered out leaving a more manageable size of litter to rear.

Recapturing Mink

The recapture of an escaped mink is best done by two or more people. One person approaches the mink as near as he dares without disturbing it, then remains absolutely still. The other(s) then attempt to run the mink towards the stationary person who captures it as it runs past him. An aid to recapturing is to use the line of the guard fence. If the mink is manoeuvred to run along the base of the guard fence, its escape options are reduced and the operation made easier

Further Reading

Blue Book of Fur Farming (Annual). Communication Marketing, Minnesota.

Joergensen, G. (1985) (ed.) *Mink Production.* Scientific, Hilleroed, Denmark.

Wilson, H. C. (1976) Mink (*Mustela vison*). In: *A Manual of the Care and Treatment of Children's and Exotic Pets* (Editor Cowie, A. F.), pp 133–146. BSAVA, London.

Index